Resilient and Friendly Living Environments

The paramount goal of this series is to be a groundbreaking resource for professionals, practitioners, and enthusiasts dedicated to shaping resilient, friendly, and inclusive environments. This series aims to empower its readers with cutting-edge insights and practical tools to design spaces that seamlessly blend sustainability, adaptability, and cultural sensitivity. By delving into such topics, the series strives to provide a holistic perspective on creating environments that prioritize the well-being and interconnectedness of diverse communities. Moreover, the series also seeks to guide the development of frameworks that evolve with the ever-changing needs of societies, ensuring that the built environment remains responsive and relevant over time. Ultimately, this book series aspires to be a catalyst for positive transformation, encouraging a paradigm shift in the way we conceptualize, design, and interact with the spaces we inhabit. This series covers a broad spectrum of topics related to building environments that stand resilient and friendly, with a special emphasis on inclusivity for individuals of all ages. The series encompasses but is not limited to:

- Sustainable and Adaptive Design: Guidance on creating environments that are environmentally sustainable and capable of adapting to changing circumstances.
- Cultural Considerations in Design: Insight into designing spaces that respect and reflect the cultural diversity of communities which promote a sense of belonging among individuals.
- Supportive Infrastructure: Detailed examinations of infrastructure elements that contribute to the overall well-being, safety, and accessibility of diverse communities. It can also include wellness initiatives that address physical, mental, and emotional well-being.
- Designing for All Ages: Foster an inclusive environment that values diversity. This includes promoting equal opportunities, and ensuring that everyone feels welcome and respected. In-depth discussions on creating infrastructure and spaces that caters to the unique needs of both children and older adults is also of great importance.
- Community Engagement: Practical approaches to involving the community in the design process and ensuring that the built environment aligns with their needs and preferences.
- Safety Measures and Disaster Risk Reduction: Prioritize safety and security to ensure a comfortable and secure environment. This includes physical safety measures as well as fostering a culture of respect and mutual support.
- Technological Integration: Explorations of how technology can be seamlessly integrated to enhance connectivity, flexibility, and inclusivity within the built environment.
- Collaborative Spaces: Design work and learning environments that facilitate collaboration. This can involve open-plan workspaces, collaborative tools, and flexible seating arrangements that encourage interaction and teamwork.
- Adaptive Policies and laws: Establish policies and laws that are adaptable to changing circumstances. This flexibility allows for the accommodation of diverse needs and ensures that policies remain relevant and effective over time.

This book series is a valuable resource for architects, designers, urban planners, educators, and anyone interested in the intersection of design, community, and well-being. The series aims to inspire and guide readers in creating environments that not only withstand challenges but also flourish as vibrant and inclusive spaces for everyone.

Yong Xiang · Yonghua Chen · Lingyun Zheng

Research on Coupling Coordination of Urban Safety Resilience Based on Disaster Prevention and Reduction

Springer

Yong Xiang
School of Architecture and Civil Engineering
Xihua University
Chengdu, Sichuan, China

Yonghua Chen
School of Architecture and Civil Engineering
Xihua University
Chengdu, Sichuan, China

Lingyun Zheng
School of Architecture and Civil Engineering
Xihua University
Chengdu, Sichuan, China

ISSN 3005-1681 ISSN 3005-169X (electronic)
Resilient and Friendly Living Environments
ISBN 978-981-96-1564-3 ISBN 978-981-96-1562-9 (eBook)
https://doi.org/10.1007/978-981-96-1562-9

This work was supported by Social Development and Social Risk Control Research Center.

This Springer imprint is published by the registered company Springer Nature Singapore Pte Ltd.
The registered company address is: 152 Beach Road, #21-01/04 Gateway East, Singapore 189721, Singapore

Introduction

Natural disasters have become more frequent as human activity and the industrial landscape have changed. These events have put unprecedented pressure and inflicted substantial losses on many urban areas. Currently, cities in many developing and even developed nations face the challenges of inadequate urban infrastructure, limited ability to handle emerging occurrences and natural disasters, inadequate disaster risk reduction (DRR) capacities, and significant environmental degradation. These challenges pose significant hurdles to urban disaster prevention and reduction management. Therefore, improving urban catastrophe preventive and reduction management systems is crucial following the ongoing urban development trend of establishing inclusive, safe, resilient, and sustainable cities.

The concept of resilient cities has emerged as a focal point in optimizing urban DRR. The academic discourse on urban resilience encompasses multiple dimensions. For instance, the Rockefeller Foundation identifies four fundamental dimensions: health and well-being, economy and society, infrastructure and ecosystems, leadership and strategy. Cities such as Kitakyushu in Japan and Hosei University propose urban indicators, administrative metrics, and civic indices as indicators. Generally, city resilience can be evaluated based on ecological, social, economic, and infrastructural-capacities aspects. Although these evaluative dimensions guide our understanding of urban resilience, safety should not be overlooked when expanding this understanding. Viewed through the lens of urban safety resilience, resilience, and safety are inextricably linked, thus drawing keen interest to urban safety resilience as a novel discourse.

Urban safety resilience theory is a novel approach to urban safety development characterized by extensive, inclusive, and adaptable analyses. This theoretical framework aids urban systems in effectively managing the uncertainties of unknown risks while adapting to continuously evolving internal and external threats. As an effective and sustainable urban development model, the strategic, proactive, and integrative nature of urban safety resilience enhances the practicality of urban systems. Consequently, assessing urban safety resilience will define new strategic directions and models for urban safety, presenting effective solutions for urban DRR management. Moreover, constructing resilient and safe cities necessitates a holistic consideration of human, infrastructural, environmental, social, and economic aspects. Achieving a balance between these elements poses significant challenges.

This book conducts a series of studies on the coupling coordination of urban safety resilience within the context of the DRR, exploring the mechanisms behind urban safety resilience from various perspectives and expanding the evaluative framework of urban safety resilience to address the inherent challenges of creating regionally differentiated safe and resilient cities. The research encompasses the Chengdu-Chongqing economic circle, constructing an urban safety resilience assessment framework through literature analysis, employing the entropy TOPSIS method to evaluate these dimensions quantitatively, and diagnosing impediments through a barrier model. Further, it utilizes an improved coupling coordination model to calculate the degree of urban safety resilience, employs kernel density estimation to determine its temporal evolution, and integrates Moran's Index using ArcGIS and GeoDA software to reveal the spatial distribution characteristics of urban safety resilience, dynamically analyzing the evolution of coupling coordination. The findings present the current state of urban safety resilience coupling and coordination in the Chengdu-Chongqing economic circle, offering genuine and reliable examples for global urban resilience development and proposing strategies from the perspective of urban sustainable development goals, thereby providing a wealth of policy recommendations for urban decision-makers.

This book is a collaborative research and writing effort by Yong Xiang, Yonghua Chen, and Lingyun Zheng of Xihua University. We incorporated insights and findings from various scholars during its creation, enriching and refining the content. We extend our heartfelt gratitude to these scholars and express special thanks to Springer Nature for their editorial support during the book's conceptualization, proposal submission, and writing process.

Figure 1 is the research framework for assessing urban safety resilience's coupling coordination degree, including five parts: (1) definition of research targets; (2) construction of an evaluation indicator system; (3) indicator weights and data sources; (4) coupling coordination analysis methods; and (5) relevant countermeasures and suggestions and conclusions.

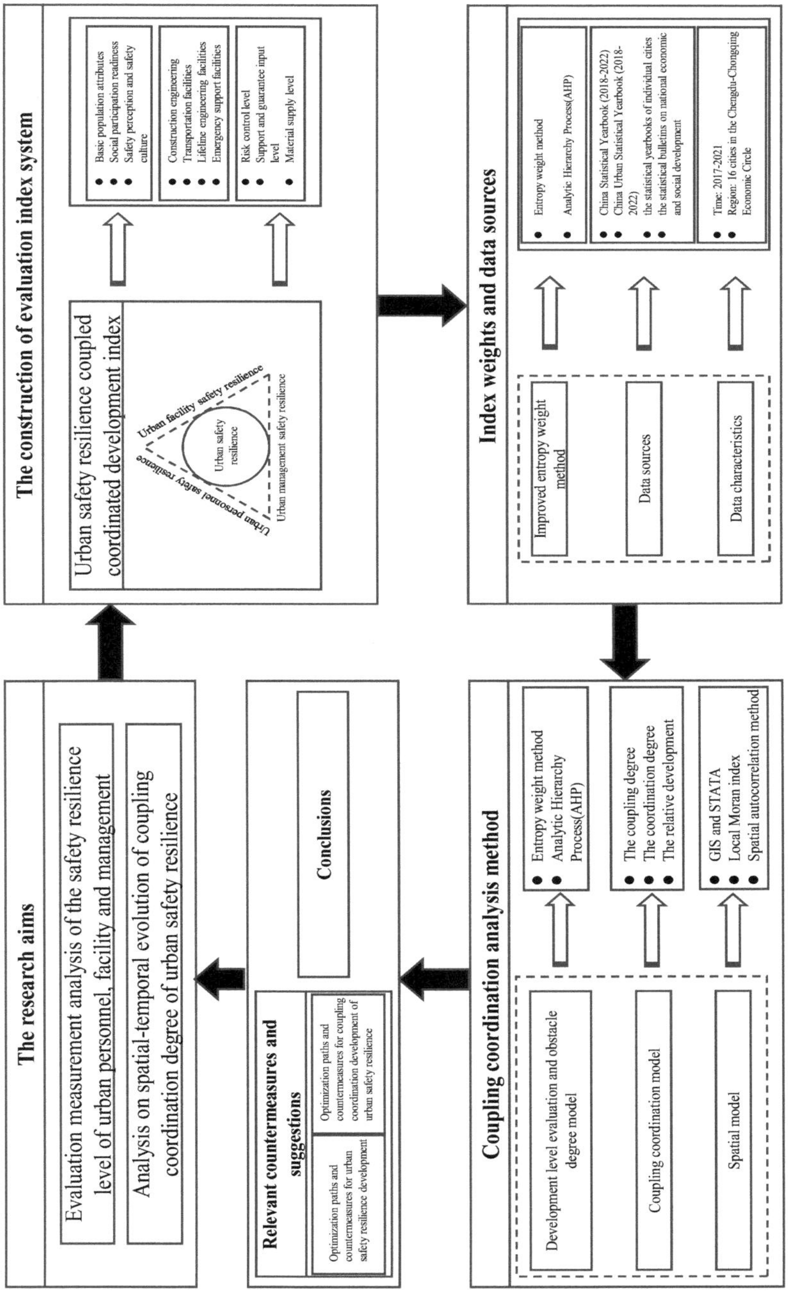

Fig. 1 Research framework

Contents

Chapter 1
Urban Safety Resilience and Disaster Prevention and Reduction

Cities serve as a nation's or region's population, wealth, and information hubs in addition to being its political, economic, and cultural hubs. In the twenty-first century, the urbanization process continues to accelerate, and the uncertainties in society, the risk factors, and disasters' inducing factors are also increasing. This characteristic of modern cities leads to disasters that will cause calamities once they occur. Accordingly, the concepts of urban disasters, disaster mitigation, and prevention are defined in this chapter. The relevant literature is compiled, and urban catastrophes' significance, effects, and implications are clarified. Meanwhile, a detailed explanation of the definition and features of urban disaster prevention and mitigation is provided.

Urban Disasters

During the process of urban development, the urban landscape changes alongside rapid economic growth and population expansion. This rapid urbanization symbolizes the advancement of modern civilization and draws people to live here in search of a higher quality of life. Moreover, numerous businesses and industries invest in the city, creating more job opportunities for its residents. Tall buildings and convenient transportation undoubtedly contribute to the city's prosperity, but unavoidable risks and challenges exist. As human habitation and life centers, cities are frequently exposed to various natural disasters. For instance, earthquakes have the potential to cause tall buildings to collapse, floods may inundate urban areas, and fires may cause more significant damage due to a higher concentration of buildings. Therefore, the danger of disasters can be increased by several factors, including rapid urbanization, population concentration, and dramatic increase in land use. We ought to comprehend these tragedies better in light of this.

Y. Xiang et al., *Research on Coupling Coordination of Urban Safety Resilience Based on Disaster Prevention and Reduction*, Resilient and Friendly Living Environments,
https://doi.org/10.1007/978-981-96-1562-9_1

Bibliometric Analysis of Urban Disasters

Keywords provide a concise document synopsis, and analyzing them can yield an initial understanding of its theme. Generally, the keywords used in a publication are inherently related. Keyword co-occurrence analysis is based on large volumes of document data, statistically correlating the keywords in each document to form a co-occurrence network (Junlong et al. 2024), as shown in Fig. 1.1.

By analyzing 100 CSSCI documents on "urban disasters" from the CNKI database, covering the period from 2000 to 2023 with yearly intervals, it is observed that the research keywords primarily focus on "disasters," such as urban waterlogging, urban crises, urban flooding, war disasters, and emergency management. Figure 1.2 presents a time zone map for keyword clustering. The time zone map clearly illustrates the stages of a research topic from inception through significant development and maturity to completion. It displays the time span and research progress of the development and evolution of each subfield (i.e., cluster) (Hong et al. 2018a). As a result, this figure offers a time zone visualization analysis of the

Fig. 1.1 Keyword co-occurrence network mapping in the CNKI database

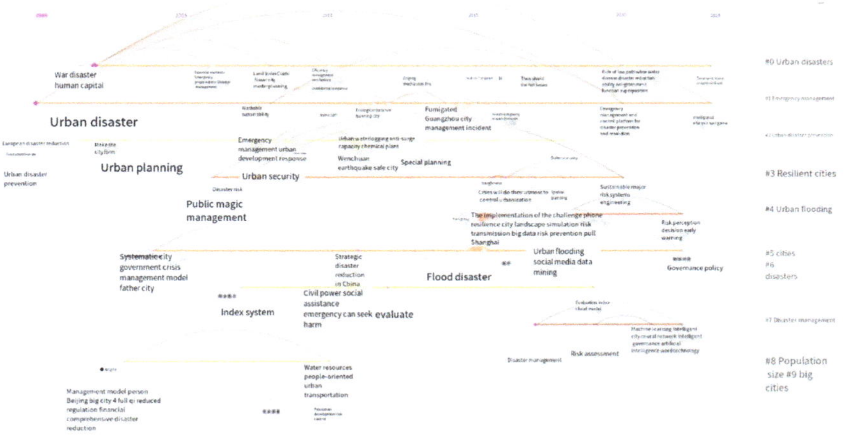

Fig. 1.2 Time zone view of urban disasters keywords in CNKI database

keywords, illustrating the evolution of temporal knowledge and shifts in research directions related to urban disasters.

According to the time zone map results, research on urban disaster-related content in the CNKI database began in 2000. Initially, the database concentrated on urban disasters and disaster prevention and mitigation. Over time, the research direction gradually shifted towards disaster management, urban resilience, and emergency response. Topics such as urban disasters, urban resilience, and emergency management have emerged as prominent trends and directions in CNKI database research.

The keyword emergent graph shows that the research is divided into four stages: From 2000 to 2005, urban disaster was in its infancy, with study methodologies mainly focused on the impact of traditional natural disasters (such as earthquakes, floods, and hurricanes) on cities (Cutter et al. 2012). Subsequently, establishing and enhancing disaster management and emergency response systems became the primary research priority. From 2006 to 2010, the research was primarily concerned with the impact of urbanization on disaster risk and the management of urban safety and emergency issues (Pelling 2007). From 2011 to 2015, the emphasis shifted to disaster assessment and comprehensive risk management, considering multiple factors such as social, economic, and environmental aspects while gradually focusing on urban resilience (Cutter et al. 2014). From 2016 to 2020, the application of digital technologies and big data in urban disaster research expanded, emphasizing climate adaptation, urban resilience, and sustainable development (Aitsi-Selmi et al. 2016). Since 2021, the emphasis has shifted to disaster governance and the adoption of interdisciplinary and multidisciplinary research methods, focusing on emerging risks (such as extreme weather events caused by climate change and public health crises) and their overall impact on cities (Sharifi and Khavarian-Garmsir 2020) as illustrated in Fig. 1.3.

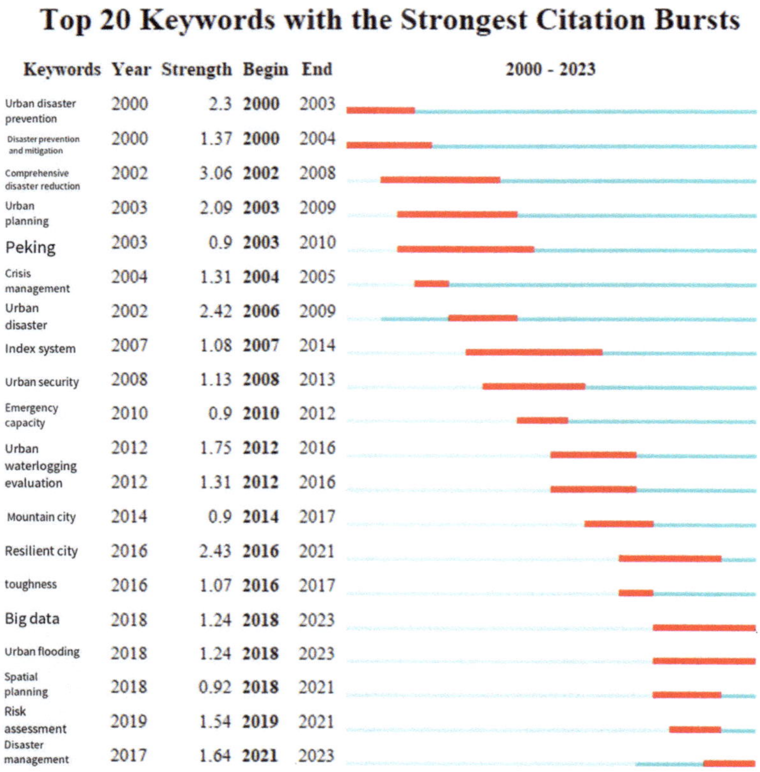

Fig. 1.3 Citation bursts of urban disasters keywords in the CNKI database

In summary, bibliometric research on urban disasters indicates a trend away from an initial focus on the types of urban disasters and their effects on cities towards adjacent topics such as urban disaster prevention and mitigation, disaster management, and urban resilience. This shift reflects not only a scholars' deepening understanding of urban disasters but also the growing need to address disaster concerns. Research methodologies have progressed from simple and limited approaches to increasingly incorporating geographic information systems (GIS) and remote sensing technologies for disaster risk assessment. Additionally, the advent of international cooperation and global perspectives has brought fresh ideas and methodologies to urban disaster research. Scholars have recently adopted comprehensive research methods to explore novel disaster management pathways and intensified efforts to study the comprehensive impacts of emerging risks on cities. Urban disaster research is trending towards multi-technical and multifaceted exploration. Consequently, the research direction on urban disasters in the CNKI database has gradually shifted towards comprehensive studies on disaster management and the impacts of multiple disaster types.

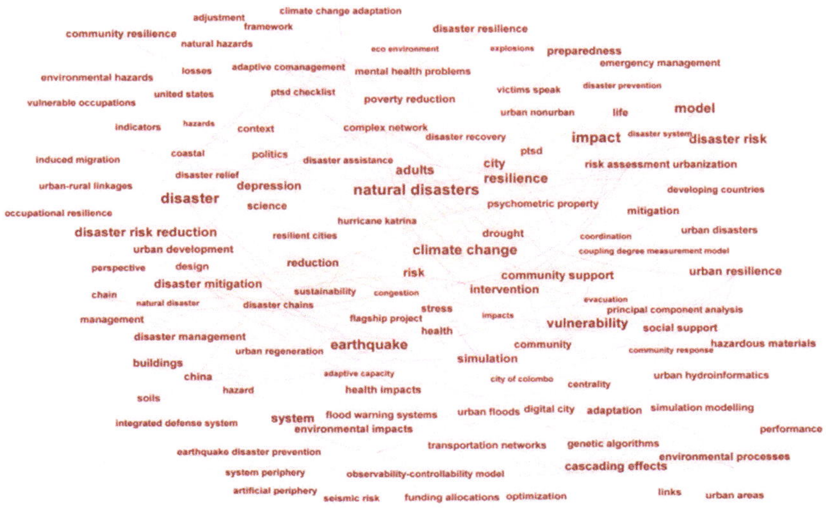

Fig. 1.4 Co-occurrence network of urban disasters keywords in the WOS database

To match the "urban disasters" research theme, the WOS database focused on fields with the highest density of interdisciplinary applications and citation rates, such as environmental science and ecology, geology, water resources, engineering, meteorology, and atmospheric science. A keyword co-occurrence network examined 147 SSCI articles on "urban disasters" from 2000 to 2023, with a one-year time slice. This network clearly depicts the dataset's structure: each node represents several publications, and the larger the node, the higher the frequency of the keyword and the more vigorous its relevance to the research theme (Huiqing et al. 2024), as illustrated in Fig. 1.4.

Prominent keywords include climate change, disaster risk, natural disasters, impact, vulnerability, and earthquake. These keywords reflect the WOS database's primary research on urban disasters, focusing on climate change, natural disasters, and disaster impacts.

Through analysis of the temporal distribution map (Fig. 1.5) of urban disasters, it is evident that in the early twenty-first century, there was a close association between emergency management in the face of disasters and natural disasters in urban areas, which occurred with relative frequency. Subsequently, keywords such as earthquakes, climate change, floods, and urban resilience have successively sparked discussions in the field of urban disasters. In 2015, urban disaster management and resilience research gradually took center stage for discussion. In recent years, disaster response has become the main focus of this field, with high frequencies of keywords such as disaster response systems and response frameworks. In contrast to the CNKI database, research on urban resilience in the CNKI database represents a new study avenue in urban disasters.

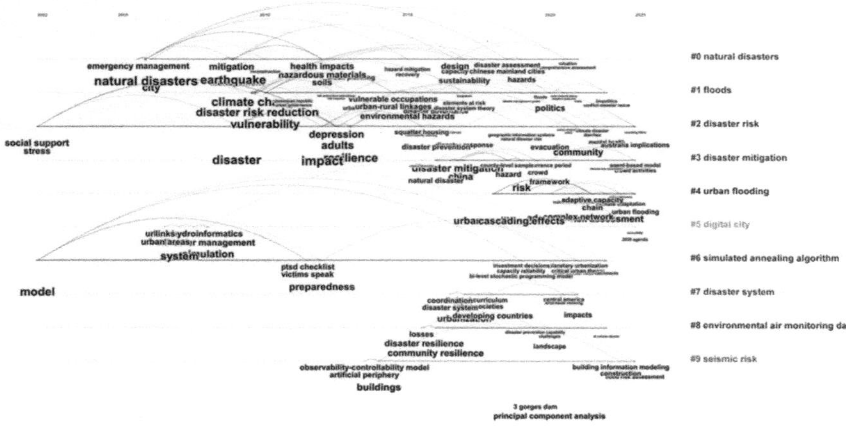

Fig. 1.5 Map of urban disaster time zones in the WOS database

Keyword emergence is divided into four stages depending on their presence over time: the early frontier stage, which lasted from 2000 to 2010, was characterized by keywords such as stress, emergency management, system, and disaster. These keywords are clearly identified to exist in the early stage of urban disasters. From 2011 to 2014, the stage of urbanization and planning was characterized by phrases such as urban planning and urban appearing frequently. The period from 2015 to 2020 represents the stage of environmental and resilience development. During this stage, emerging words such as environmental hazards and disaster resilience reflect an emphasis on environmental risk assessment and management, urban resilience development, disaster mitigation measures, and the impact of climate change on disaster management. Since 2020, the focus has switched to systemic frameworks and community resilience. Keywords such as framework and community are prevalent during this period, highlighting comprehensive disaster mitigation strategies, systemic frameworks, community resilience, and the practical implications of disaster management as the primary research focuses. See Fig. 1.6 for visual illustration.

The literature on urban disasters in the Web of Science database shows that research on natural disasters and urban resilience has progressed from emergency management to comprehensive systemic approaches. Early research primarily focused on establishing emergency response and management mechanisms, then shifted towards urban planning and risk management, further addressing environmental impacts and resilience building, eventually evolving into comprehensive systemic frameworks and community resilience enhancement. This evolution reflects the deepening and broadening of disaster management research as urbanization

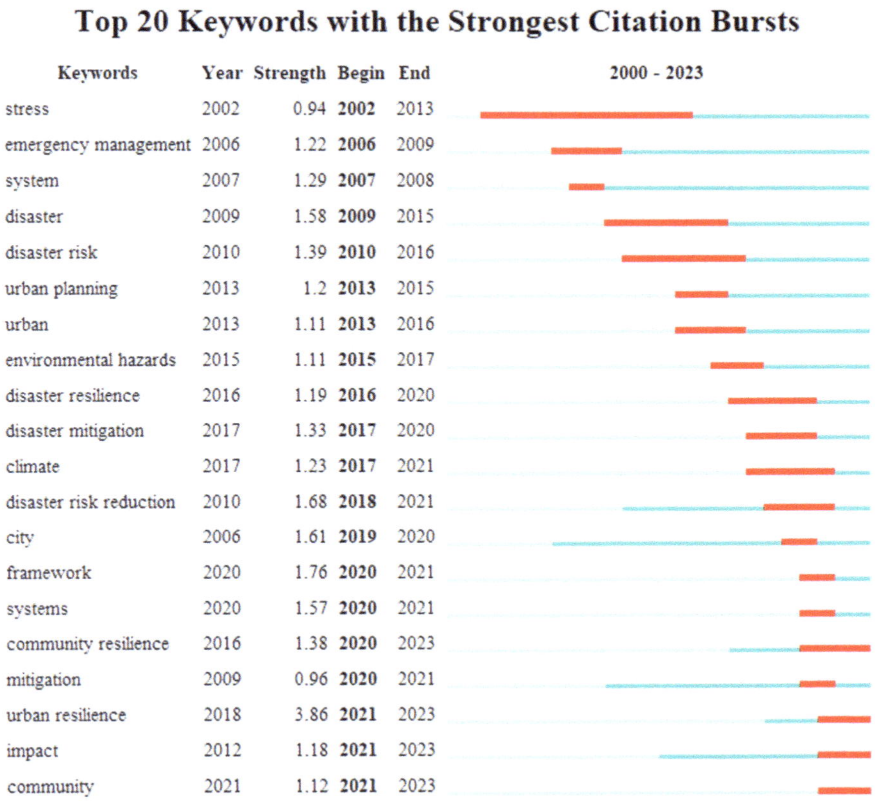

Fig. 1.6 Urban disaster emergent terms in the WOS database

progresses and the impacts of climate change intensify, thereby addressing increasingly complex disaster risks and challenges. Future research will emphasize interdisciplinary collaboration and the implementation of comprehensive strategies to improve the overall effectiveness of disaster management and the practical effects of resilience building.

Literature Review on Urban Disasters

The frequent occurrence of natural disasters inflicts significant losses on the urban economy, society, and individuals' lives and property. The acceleration of urbanization and exacerbation of climate change underscores the growing importance of urban disaster research. According to the analysis of two databases in the preceding sections, research on urban disasters can be categorized into the following aspects:

Research on different types of disasters and their effects: Early urban disaster research primarily focused on the types of urban disasters and their direct impacts on cities. Scholars analyzed the impacts of natural disasters (e.g., earthquakes, floods, and typhoons) and man-made disasters (e.g., fires and explosions) on urban infrastructure, economy, and society using qualitative and quantitative methods. L. Zhang et al. combined AHP, F-AHP, and AHP-TOPSIS to assess the geological disaster risk in the target area, revealing that slope debris and rainfall are the primary factors leading to disaster occurrence (Zhang et al. 2022). Birgani, Y. T. et al. studied the impact of flood disasters on different types of land, buildings and the damage to citizens' property and lives (Tahmasebi Birgani and Yazdandoost 2018). Chen, W. F. et al. assessed the typhoon disaster risk in the Yangtze River Delta region using principal component analysis (Wenfang et al. 2011). Hallegatte and Dumas' research indicates that natural disasters are generally adverse exogenous factors for regional economic development, particularly large-scale and significant natural disasters often deal a fatal blow to local industrial development and human capital, even leading economically fragile areas back into poverty (Hallegatte and Dumas 2009). Thomas examined the welfare losses of the Vietnamese people from a micro-perspective and discovered that hurricanes and floods caused the highest welfare losses, thus supporting the conclusion that natural disasters can disrupt local economic growth and development at the micro-level (Thomas et al. 2010).

Therefore, the impacts of single or multiple disasters on cities are multifaceted. However, as research progresses, scholars have gradually realized that merely understanding the types and impacts of disasters is insufficient to address urban disasters effectively. Consequently, research efforts have shifted towards establishing and improving urban disaster prevention and mitigation mechanisms, as well as disaster governance systems. By exploring more systematic and comprehensive disaster prevention and mitigation strategies, enhancing emergency response capabilities, and formulating effective disaster governance policies, cities can better protect themselves from disaster damage and enhance their resilience in disasters.

Disaster management aspects: It is well known that the government and relevant departments, military forces, and social forces (i.e., non-governmental organizations, enterprises, the public, and mass media) are the three main entities in disaster response, constituting the primary components of a multi-subject collaborative governance system in disaster response. Therefore, these three entities need to be organically integrated in disaster governance. Zhang Q. et al. pointed out that for successful and orderly disaster governance, multiple entities such as government, social organizations, enterprises, and the public must participate in an orderly manner (Qin et al. 2010). Qi X. L. et al. recommended that strategies for disaster governance should start by enhancing the public's emergency awareness, establishing and improving basic disaster laws, sound scientific emergency plans, and enhancing post-disaster information communication channels (Xiaoliang et al. 2019). Zeng G. P. et al. analyzed the model of natural disaster response with the government as the main body and multiple entities participating. They found that efforts should be made to strengthen the system of natural disaster emergency plans, establish corresponding laws and regulations for disaster response, conduct routine simulation exercises for emergency rescue, improve the allocation and use of disaster relief supplies, establish multi-level professional emergency rescue teams, enhance the mechanism for multiple participation in disaster relief processes, and establish emergency management research platforms, thus strengthening the disaster response system (Guoping et al. 2010). Li L. pointed out the importance of clarifying the roles of government and non-governmental organizations in public crisis management for effective governance (Li 2016).

Collaborative efforts among government, businesses, and communities to formulate effective governance policies and enhance emergency response capabilities by establishing a systematic disaster management framework enable cities to handle and reduce disaster losses better. However, relying solely on policy and mechanism improvements is insufficient to cope with the complex and evolving disaster risks. Therefore, the application of modern technological methods is becoming increasingly crucial.

The continuous advancement of technology has played a significant role in urban disaster research, with geographic information systems (GIS) and remote sensing technology. First, GIS technology can provide precise geographic data and disaster simulations, aiding in spatial analysis of risks and management decisions. As an illustration, Yang J. et al. developed a business system utilizing VB.NET in the ArcGIS Explorer development environment, which is designed for displaying, processing, and analyzing various meteorological elements, assisting in preventive meteorological disaster services (Jun et al. 2013). Second, remote sensing technology enables real-time monitoring and assessment of dynamic changes in disasters, providing strong support for disaster early warning and emergency response. Wu, H. J. et al. proposed analyzing the relationship between building damage degree, material structure, and earthquake casualties through remote sensing imagery and GIS data analysis. They developed a model for estimating earthquake casualties based on remote sensing impact (Hengjing et al. 2013). Dell'Acqua et al. created the Global Exposure Database for the Global Earthquake Model (GED4GEM) database required for building a global earthquake model. They examined the advantages and limitations of remote sensing data and spatial analysis technology in refining exposure data processing and analyzing the application potential of remote sensing data (Dell'Acqua et al. 2013).

Development of technical means: Advanced technologies such as GIS and remote sensing have significantly improved the accuracy and efficiency of disaster management. However, when facing increasingly complex urban disaster risks, relying solely on a single technological approach remains inadequate. Therefore, the introduction and application of integrated systematic methods become increasingly important.

Integrated systematic approaches and urban resilience development: The evolution from emergency management to integrated systematic methods signifies the deepening of urban disaster research. The research emphasis has transitioned from emergency response and management mechanisms to urban planning and risk management and subsequently to environmental impact assessment and urban resilience development. Current research increasingly adopts integrated systematic frameworks, focusing on enhancing resilience at the community level to achieve more comprehensive and effective disaster management. Wang et al. believe that integrated disaster management is critical to urban sustainable development. They employed knowledge graph analysis to examine the development of integrated disaster management, providing a concise overview of cutting-edge research worldwide. The analysis focused on climate change, public health, infrastructure, insurance, communities, and other social organizations (Wang et al. 2021). Minxing Yang et al. conducted a retrospective review of disaster prevention and control research, considering resilience city theory as a new approach for urban disaster prevention and control (Minxing et al. 2016). Hence, establishing resilient cities is vital to disaster risk management (Hui et al. 2020).

In conclusion, urban disaster research tends to explore multiple technologies and aspects. From early research focusing on disaster types and their impacts to modern comprehensive studies on disaster prevention, mitigation, governance, and urban resilience, this process reflects scholars' deepened understanding of urban disasters and the demand for addressing disaster challenges. Future research will focus more on interdisciplinary collaboration and the implementation of comprehensive strategies to further enhance the overall effectiveness of disaster management and the construction effect of urban resilience. Facing increasingly complex disaster risks and environmental changes, scholars need to integrate multidisciplinary knowledge, apply advanced technologies, and implement multi-level strategies. The research should include enhancing international cooperation, sharing and acquiring knowledge from global best practices; exploring the comprehensive impact of emerging risks such as climate change and urbanization on cities; and developing innovative disaster governance methods and resilience-building strategies. Through interdisciplinary collaboration and systematic research, urban disaster management will be more comprehensive and efficient, significantly enhancing cities' resilience and recovery capabilities.

Context of Urban Disasters

1. Definition of Disaster

The definition of disaster has undergone a complex development process, evolving from the initial concept of natural disasters caused by earth, fire, and water to encompass various scenarios (Yongxiang 2011). Currently, there are many different interpretations regarding the definition of disaster. In China, the interpretation of disaster is based on the "Cihai" (7th edition), which defines it as damage caused by natural and human-induced disasters. On a global scale, the Glossary of Terms for Disaster Health Risk Management from the World Health Organization's Glossary is widely recognized. A hazardous event can cause severe disruption to the functioning of a community or society at all scales. This is caused by the interplay between exposures, vulnerabilities, and capacity conditions, resulting in one or more of the following consequences: human, material, economic, and environmental losses and impacts. Scholars from different disciplines have different interpretations of disasters (Table 1.1).

Disasters are interpreted differently across various disciplines. However, they are typically defined as sudden events or situations, whether caused by nature, human actions, or their interaction, that severely and persistently affect human society, economy, and environment, surpassing the coping capacity of a given community or society. An urban disaster is thus a potential danger that threatens the urban population and associated economic activities (Yongxiang 2011).

2. Causes of Disasters

The urban natural disaster system is an essential component of Earth's intricate megasystem. The onset and progression of natural disasters are shaped by factors such as Earth's processes, solar activity, and material fluctuations across various spheres. Simultaneously, it is also influenced by social factors such as urban population, resources, environmental changes, social, and economic development. Therefore, it can be assumed that the generation of the urban natural disaster system originates from the interaction between variations in the earth system and changes in the social system. This system is a subsystem of the complex giant system of heaven, earth, human, society, and earth (Yongxiang 2011). Thus, the causes of urban disasters can be categorized into three parts: disaster-causing factors, disaster-induced environments, and disaster-bearing bodies.

The disaster-causing factors are considered the primary causes of urban disasters, including earthquakes, floods, hurricanes, heavy rains, and similar events that often cause significant damage and casualties. The geographical characteristics of the city, such as earthquake zones and flood-prone areas, also directly influence the potential risk of natural disasters. Additionally, human activities significantly contribute to urban disasters, including industrial accidents, chemical leaks, and traffic accidents, which can have severe consequences in cities. Inadequate urban planning, environmental pollution, and poorly designed buildings can also increase the frequency of disasters. A single factor can lead to a disaster, and combining multiple factors into a composite factor can make the disaster even more devastating. For instance, earthquakes can trigger landslides, floods can cause mudslides, and the combination of these factors can increase the complexity and destructive force of the disaster.

Given the emergence of factors leading to disasters, it is essential to analyze the specific environment in which these factors arise. It is because of the direct correlation between the environment in which disasters occur and the factors causing them.

Table 1.1 Definition of the disaster from different disciplines

Academic disciplines	Definition of disaster
Economics	Disasters are essentially about economic loss
Sociology	Disaster is a social structure and social process
History	Disasters are a history of human struggle against disasters
Anthropology	The core of disaster is the cultural construction of disaster

For example, aging, imperfect, or disaster-unfriendly infrastructure increases the incidence of urban disasters. Additionally, unstable water conservancy projects and weak building structures make cities less able to cope with disasters. Cities built on mountains are characterized by landslides due to earthquakes. Mudslides can wipe out cities at the bottom of basins, and cities emerging along the coast are vulnerable to storm surges and tsunamis.

The disaster-bearing body carries the disaster-causing factors in the vulnerable environment. It encompasses individuals, communities, and organizations who experience the effects of the disaster and are responsible for personally handling and enduring its consequences. Its presence reflects the magnitude of the disaster's impact.

Therefore, disaster-causing factors, environments, and affected entities are interconnected, shaping the origins of urban disasters. Figure 1.7 (Guijuan 2011) illustrates the formation mechanism of urban natural disasters.

3. Types of Disasters

Disasters in cities are inevitable, but it is vital to comprehend their nature and classification. Researchers have examined urban disasters and classified them into 11 categories: urban earthquake disasters, urban flood disasters, urban meteorological disasters, urban geological disasters, major infectious diseases, fires and explosions, urban lifeline system accidents, major industrial accidents, urban environmental pollution hazards, terrorist attacks, and information security disasters (Xiaoya 2012). Additionally, other researchers classify urban disasters into three primary categories based on their causes: natural disasters, technology disasters, and social disasters

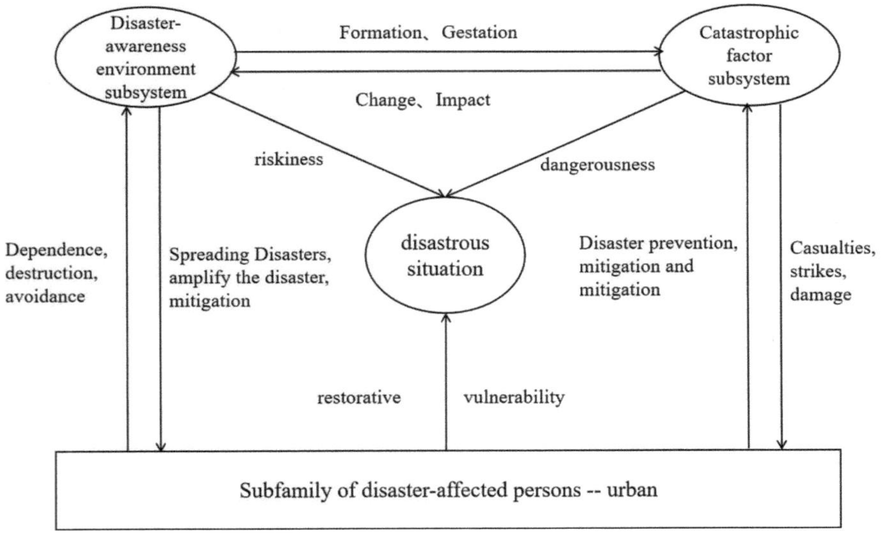

Fig. 1.7 Mechanisms of urban disaster formation

(Bo 2014). Moreover, researchers have concluded that urban system disasters can be broadly categorized into two main categories: natural man-made disasters (e.g., earthquakes, tsunamis, storm surges, typhoons, tornadoes, saltwater intrusion, coastal erosion, mudslides, floods, ground settlement and subsidence, landslides, and fires,) and construction safety-related disasters (e.g., partial and total collapse of houses, fatalities, and injuries, hazards in underground construction, and accidents related to pits and subway construction, as well as incidents in municipal road pipeline network operations such as pipeline rupture, explosion, gas and natural gas leakage) (Bo 2008). Different researchers have employed diverse categorization methods for urban disasters.

(1) By Causes:

① Natural disasters: Natural disasters are caused by natural forces, such as earthquakes, tsunamis, floods, and typhoons.
② Man-made disasters: Man-made disasters are caused by human activities, such as industrial accidents, transportation accidents, and terrorist attacks.

(2) By Occurrence Frequency:

① Conventional disasters: Conventional disasters generally occur with high frequency, such as traffic accidents and fire.
② Rare disasters: Rare disasters, such as earthquakes and floods, occur with low frequency.

(3) By Disaster Form:

① Meteorological disasters: Meteorological disasters are related to weather and atmospheric conditions, such as heavy rain, snowstorms, and hurricanes.
② Geological disasters: Geological disasters are related to geological processes like mudslides and volcanoes.
③ Hydrological disasters: Hydrological disasters involve water-related events, such as floods, droughts, and tsunamis.
④ Biological disasters: Biological disasters involve biological agents, such as infectious diseases and pest outbreaks.

(4) By Impact Magnitude:

① Regional disasters: Regional disasters have localized impacts on a specific area or region, such as roads collapsing and houses collapsing.
② Global disasters: Global disasters have widespread impacts that cross national and continental boundaries, such as rising sea levels and global epidemics.

(5) By Development Stage:

① Predictive disasters: Predictive disasters can be forecasted with greater accuracy in advance, such as typhoons and rainstorms.

② Sudden onset disasters: Sudden onset disasters occur suddenly and are difficult to predict, such as explosions.

Therefore, the classification of urban disasters is interrelated, and there are overlapping situations with various classification criteria.

4. The Characterization of Disasters

Categorizing disasters is the foundation for comprehending and investigating various types of disasters influenced by multiple direct and indirect factors. This complexity makes the study of urban disasters particularly intriguing. The characteristics of urban disasters are as follows:

(1) Hazardous Nature: First, disasters present a direct threat to human life. The high dense urban population and extensive infrastructure exacerbate this risk, potentially leading to widespread casualties and irreversible humanitarian crises. Second, disasters cause extensive destruction to infrastructure. Crucial elements involving transportation, energy, water supply, and drainage are vital for the proper functioning of urban areas, and their collapse due to disasters can throw the city into disarray. Moreover, disasters have enduring economic and social repercussions on cities, including business closures and rising unemployment rates, thus jeopardizing the cities' economic sustainability.
(2) Diversity: Urban disasters exhibit greater diversity compared to those in other regions. Due to variations in geographic locations, environments, climates, economies, cultures, and other factors among different cities, the causes, forms, and impact magnitude vary significantly. With high population density, extensive buildings, and diverse economic activities, urban areas are exposed to numerous disaster-causing hazards and a wide range of disaster types (Wenqian 2020). Regarding natural disasters, cities may face geological hazards (e.g., earthquakes), meteorological hazards (e.g., hurricanes), hydrological hazards (e.g., floods), and biological hazards (e.g., epidemics). Each of these risks has its distinct mechanism of occurrence and impact. Man-made disasters, such as industrial accidents, transportation incidents, and terrorist attacks, are caused directly by human activities and exhibit distinct characteristics from natural disasters. The combination of these factors, such as earthquakes triggering fires and hurricanes causing tsunamis and floods, leads to diverse disasters.
(3) Immediacy: Immediacy is primarily evident in the suddenness and unpredictability of disasters. Many natural disasters (e.g., earthquakes and hurricanes) and sudden man-made disasters (e.g., terrorist attacks) often occur abruptly and cannot be accurately forecasted in advance. Additionally, concerning urban residents' needs and coping strategies, immediate requirements include shelter, food, water, and emergency medical assistance following a disaster. The city's emergency response mechanisms must promptly address these fundamental needs to alleviate the disaster's impact on individuals.
(4) Chain Dynamics: Certain disasters exhibit cyclical patterns. Considering the unique urban landscape, the recurring disasters initiate a cascade effect that may

result in secondary and simultaneous occurrences of multiple disaster types. In locations with dense urban infrastructure, remnants of older building complexes continue to exist, and local earthquakes frequently trigger secondary disasters of fires. Moreover, sudden onset disasters (e.g., earthquakes and floods) may induce ground subsidence, leading to substantial damage to the structural integrity of cities. The interconnectedness escalates the complexity and immediacy of urban disasters.

Urban disasters exhibit multifaceted and intricate attributes, providing a complicated setting for disaster-prone cities. A thorough comprehension of these traits can aid cities' scientific and efficient planning, enhancing response capabilities, and mitigating potential risks. Consequently, this serves as a crucial reference for fostering sustainable urban development.

Impacts and Mechanisms of Urban Disasters

1. Impact of Disasters

Cities play a central role in human civilization and symbolize prosperity and vitality in our daily lives. Nevertheless, as urbanization progresses rapidly, cities face a multitude of natural and anthropogenic disasters. These impacts not only directly threaten the safety of urban residents' lives and property but also profoundly affect urban infrastructure, economic development, and social stability. Hence, understanding and analyzing these effects is crucial for effectively managing urban disasters. The following section will clarify the impacts of disasters from a causal perspective.

(1) Natural Impacts:

Natural hazards can have particularly severe repercussions in urban environments. Due to the high population density and advanced developed infrastructure, cities are more vulnerable to disasters.

Earthquakes are prevalent natural disasters, particularly in urban areas situated near seismically active zones. Their destructive force stems from the movement of tectonic plates and the subsequent release of energy, resulting in severe outcomes such as building collapses, road destruction, and pipe ruptures. Following an earthquake, cities may suffer from casualties, interruptions in water supply, and power outages, significantly impacting urban operations and residents' livelihoods. Hurricanes and typhoons represent significant natural disasters encountered by urban areas. The powerful gusts and heavy rainfall associated with these extreme storms can result in flooding, leading to structural damage, road closures, and electrical outages. High-rise and densely populated urban areas may exhibit heightened vulnerability to hurricanes, whereas coastal cities risk devastating storm surges, resulting in extensive flooding and coastal damage. Besides earthquakes and hurricanes, flooding represents another prevalent natural disaster in urban areas. Floods can result from intense precipitation, snowmelt, river overflow, or dam breaches. Urban flooding can lead to issues like clogged sewers, flooded basements, and transportation interruptions, significantly impacting urban infrastructure and residents' daily lives. Especially in highly urbanized regions, urban drainage systems may fail to manage water drainage adequately, exacerbating the risk and consequences of flooding. Moreover, volcanic eruptions, tornadoes, and heavy rainfall constitute additional natural disasters that can profoundly affect urban areas. Volcanic eruptions can lead to the emission of ash, lava, and pyroclastic debris, adversely impacting transportation, air quality, and agricultural activities within the urban vicinity. Tornadoes can generate powerful winds and swirling vortexes, leading to structural collapses, tree destruction, and other associated ramifications. Intense rainfall can precipitate urban flooding and landslides, increasing the vulnerability and consequences of disasters in urban areas.

(2) Man-made Impacts

Man-made impacts refer to various facets of urban planning, construction, management, and environmental protection. These impacts can be as severe as, or even worse than, natural disasters.

Irrational urban planning and construction are major factors in the occurrence of man-made disasters. During urbanization, certain cities may experience disorganized planning and improper land use, leading to inadequate urban infrastructure and vulnerability to natural disasters. Urban areas may exacerbate the impacts of natural disasters due to characteristics including high-density construction, congested roadways, and inadequate drainage systems. Moreover, certain cities may be located in regions prone to natural disasters, such as flood-prone and earthquake-prone zones, elevating the likelihood of disaster exposure. Environmental pollution and resource depletion constitute significant factors contributing to man-made disasters. As industrialization and urbanization accelerate, cities may experience a surge in pollutants, such as industrial emissions, wastewater discharges, solid waste, and other contaminants, resulting in severe ecological damage. Environmental pollution has a negative effect on residents' quality of life and can lead to a range health problems, such as respiratory and skin ailments. Excessive utilization and mismanagement of resources can result in the exhaustion of resources and environmental degradation, increasing the likelihood of urban disasters. Traffic accidents and safety concerns in urban areas also contribute significantly to the occurrence of man-made disasters. In addition, cities may harbor various safety hazards, encompassing building, food, and public safety concerns, which could precipitate safety incidents and disasters, negatively impacting urban stability and growth. Inadequate urban management and corruption may exacerbate the impact of urban disasters. Certain cities may grapple with unauthorized construction, illicit mining activities, and unlawful sewage disposal, resulting in environmental degradation, resource depletion, and heightened disaster vulnerability. Human factors may worsen the social and economic challenges within urban areas. Social disparities and polarization between socio-economic classes in urban settings can engender recurrent social tensions and conflicts. Specific cities may experience economic downturns and employment challenges, posing risks to urban development and stability. These societal challenges and economic burdens can compound the effects of urban disasters, amplifying cities' vulnerability and resilience.

Cities must implement effective measures to enhance urban planning and management, enhance environmental quality, bolster urban resilience to disasters, and mitigate the damage and impact of man-made disasters.

2. Effects of Disasters

Disaster effects encompass a disaster's direct and indirect impacts on the affected and neighboring areas, spanning human, economic, social, environmental, and other dimensions. These effects can have short- and long-term consequences, underscoring their significance for the disaster area and post-disaster reconstruction efforts. Here are some common disaster effects.

(1) Human Effects

Human casualties are one of the most immediate effects of disasters. The number and severity of casualties depend on factors such as the disaster's type, scale, location, population density, and disaster preparedness measures in place. Natural disasters like earthquakes, floods, and storms often injure many individuals, particularly in densely populated urban regions. Similarly, man-made disasters like traffic accidents, industrial incidents, and terrorist attacks can also result in severe human casualties. Casualties not only directly impact the safety of the affected population but also cause family disruptions and social unrest. Disasters also affect human life, property, and living conditions. Especially during the later stages of a disaster, life and property security challenges in affected areas may persist, necessitating long-term rescue and reconstruction endeavors. Disasters affect life and property security, affecting individuals and families and the broader economy, exacerbating social security issues, and causing social unrest and instability. Disaster events can also severely impact people's mental health, leading to panic, anxiety, and depression.

(2) Chain Effect

The chain effect occurs when one event initiates a series of interconnected occurrences that magnify and intensify its impact. The chain effect typically denotes a series of secondary or subsequent urban disasters instigated by a primary disaster event, which extends its influence. Urban natural disasters may cause a chain-effect reaction. An earthquake may induce a building collapse, subsequently igniting a fire that spreads and worsens the damage. Floods may also precipitate secondary hazards, such as earthquakes and landslides, further exacerbating the magnitude of flood-related damage. These secondary disasters broaden the scope and magnitude of the original disaster's impact, making response efforts harder. Urban disasters can initiate a cascade of social repercussions. A severe flood might disrupt road transportation, hinder rescue and material transport efforts, and consequently impact the lives of affected residents. Economic hardships and social instability following a disaster may cause social unrest, compounding the challenges of post-disaster recovery and reconstruction. Damage to buildings and infrastructure after a disaster may disrupt productive and commercial activities, impacting the cities' overall economic functioning. Moreover, post-disaster reconstruction and rehabilitation endeavors may demand substantial financial resources, adversely impacting the cities' economy.

(3) Magnification Effect

The amplification effect, which refers to a scenario where a minor initial change or event triggers a series of subsequent reactions, leading to a transformation that significantly surpasses the magnitude of the original change, is particularly relevant in urban disaster situations. In such situations, localized disaster impacts propagate and intensify throughout the urban infrastructure. Additionally, social factors can induce magnification effects. For instance, a flood of a particular magnitude might incite social panic and reckless behaviors, potentially leading to broader casualties and property loss due to impulsive actions. Furthermore, panic may trigger social unrest, exacerbating the disorderly conditions in the disaster aftermath. Economic factors can also magnify the impact of disasters, disrupt business operations, halt production activities, and result in higher unemployment rates. This economic escalation may exacerbate social tensions and instability post-disaster. Strengthening urban resilience and post-disaster-coping capacity, which involves enhancing infrastructure development, improving disaster early warning mechanisms, and establishing effective rescue and recovery systems, are crucial strategies for mitigating the impact of the amplification effect.

Urban Disaster Prevention and Reduction

Bibliometric Analysis of Urban Disaster Prevention and Reduction

An analysis of 55 CSSCI papers on "urban disaster prevention and reduction" from the CNKI database was conducted. The papers were selected from 2000 to 2022 using a yearly time slice. The analysis revealed that the research keywords in the CNKI database mainly focus on "disaster prevention and reduction." These keywords include climate change, emergency management, urban safety, disaster prevention and reduction, and emergency management, as depicted in Fig. 1.8.

Based on the time zone map of urban disaster prevention and reduction, it is evident that research on this topic began quite early. From a temporal perspective, disaster prevention and reduction articles have appeared since 2000. Early research primarily focused on the prevention of urban floods and earthquakes. Over time, research gradually expanded to encompass comprehensive urban disaster reduction, including disaster reduction planning and establishing disaster prevention systems. During this period, the focus of urban disaster prevention and reduction research gradually shifted. At first, the research focused on specific disaster events, such as preventing and responding to floods and earthquakes. As research deepened,

Fig. 1.8 Co-occurrence network of keywords in CNKI database

the attention turned towards more comprehensive strategies for disaster reduction, including urban disaster reduction planning, systematic disaster prevention measures, and multi-hazard collaborative governance. Urban safety gradually became a central research issue, emphasizing the protection of urban residents and infrastructure during disaster prevention and reduction processes. Recent research trends indicate that with technological advancements, the application of artificial intelligence in urban disaster prevention and reduction has emerged as a new research hotspot. Researchers have begun exploring the application of artificial intelligence technology to improve urban resilience and enhance capabilities for intelligent disaster prevention and reduction. Artificial intelligence has significantly improved the efficiency and accuracy of urban disaster responses through its utilization in disaster prediction, real-time monitoring, and emergency response. Overall, research on urban disaster prevention and reduction in the CNKI database has evolved from single-disaster prevention to comprehensive disaster reduction and urban safety, and ultimately to the application of artificial intelligence. Urban disaster prevention and reduction research will continue to evolve towards intelligent and diversified directions, with the integration of urban safety and artificial intelligence technology emerging as a prominent trend. The time zone map of keywords clustering is shown in Fig. 1.9.

The keyword emergence map reveals that prominent keywords include "comprehensive disaster reduction," "urban disaster prevention," "resilient cities," and "disaster reduction planning." These keywords reflect the fundamental concerns and evolutionary trajectory of urban disaster prevention and reduction research. Overall, the research can be roughly divided into three distinct stages. The first stage, spanning from 2000 to 2005, primarily aimed to raise awareness and preliminary exploration

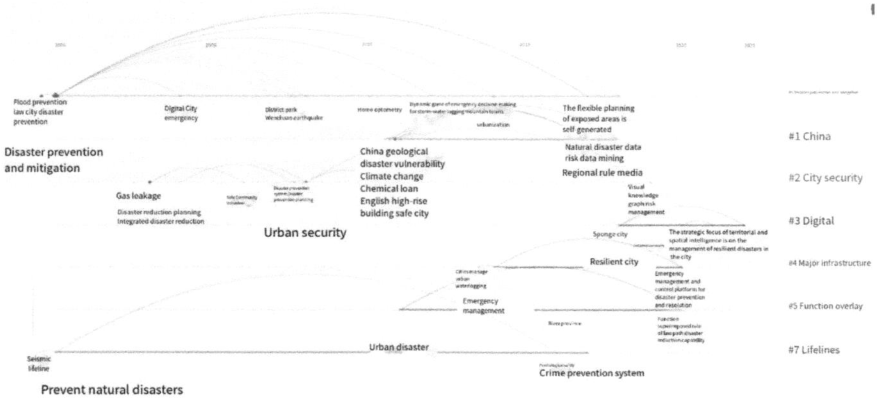

Fig. 1.9 Time zone view of urban disaster prevention and reduction keywords in CNKI database

of disaster prevention and reduction. The second stage, occurring between 2006 and 2010, witnessed a shift towards comprehensive disaster prevention and emergency management. This involved activities such as disaster planning and establishing disaster prevention systems. The third stage, starting from 2011 until the present, has moved towards multidisciplinary and multi-field integration. This stage emphasizes combining technology and management and the construction of resilient cities and urban safety, with an increased focus on social participation. The emergence of keywords such as "resilient cities," "urban safety," and "resilience" indicate that these research themes have gradually become hotspots and trends in recent years. This reflects the shift in urban disaster prevention and reduction research from addressing single disaster events to developing comprehensive, systematic, and intelligent approaches, emphasizing the enhancement of overall disaster resistance and resilience of cities through various technological means and social participation. The citation bursts of urban disasters keywords is shown in Fig. 1.10.

In summary, bibliometric analysis reveals that research on urban disaster prevention and reduction in the CNKI database has significantly developed and transformed. Originally centered on urban disaster prevention, comprehensive disaster reduction, and disaster reduction planning, research has progressively expanded to more systematic and diverse fields, including comprehensive disaster prevention planning, disaster prevention system establishment, and urban resilience. This shift signifies a change in the research emphasis, moving away from focusing on disaster response and towards systematic disaster prevention planning and multi-hazard integrated management. This evolution demonstrates the progression of research methods and viewpoints, and highlights the growing emphasis on urban comprehensive risk resistance and overall safety levels. Additionally, urban safety and resilience investigation has recently gained significant attention. Urban safety and resilience refer to a city's adaptability and recovery capacity in the face of disasters and its ability to resume normal operations post-disaster quickly. Research indicates that enhancing

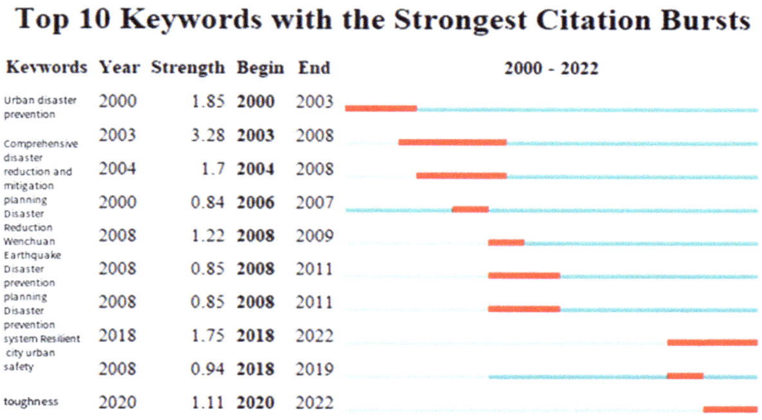

Fig. 1.10 Citation bursts of urban disaster prevention and reduction keywords in the CNKI database

urban safety and resilience decreases direct losses and promotes long-term sustainable urban development. As a result, studies on improving urban resilience through planning, infrastructure construction, social governance, and the application of new technologies have become a focus for academics and policymakers. In the future, integrating urban disaster prevention and reduction with resilient city construction will become increasingly significant. This integration will not only bolster the overall resilience of cities to risks but also provide critical theoretical and practical support for sustainable urban development.

We conducted statistical and analytical studies on urban disaster prevention and reduction by analyzing 149 SSCI papers from the WOS database. The analysis covered the period from 2006 to 2023 using one-year time slices. The findings show significant development and evolution in urban disaster prevention and reduction research during the 17 years period. Figure 1.11 presents a keyword network diagram of the Social Sciences Citation Index (SSCI) related to urban disaster prevention and reduction. The network analysis of these keywords focuses on several core topics: "climate change," "sustainability," "earthquake," "community resilience," and "vulnerability." Urban disaster prevention and reduction research involves a wide range of aspects, covering macro-level strategies for addressing climate change and promoting sustainable development, and micro-level considerations such as community resilience and individual vulnerability.

Upon examining the time zone map of urban disaster prevention and reduction in Fig. 1.12, it is evident that between 2006 and 2010, the main emphasis of research was on risk management of urban disasters. However, the research focus has shifted towards other directions since 2010, such as types of disasters, disaster prevention, and urban resilience. Urban disaster prevention and reduction research has witnessed a growing utilization of technology in recent years. This includes adopting advanced tools such as geographic information systems, the Internet of Things, remote sensing technology, and artificial intelligence.

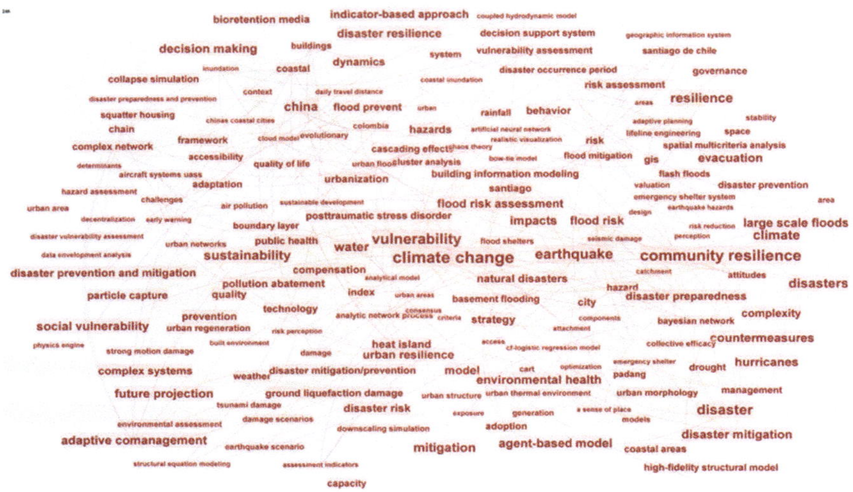

Fig. 1.11 Co-occurrence network of urban disasters prevention and reduction keywords in the WOS database

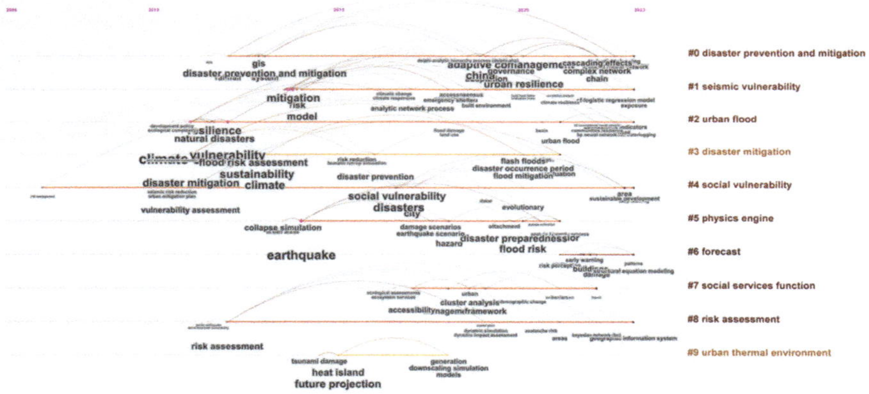

Fig. 1.12 Map of urban disaster prevention and reduction time zones in the WOS database

During the period from 2014 to 2017, the keyword burst analysis revealed that the keywords “risk” and “reduction” had higher intensity. This indicates an increased focus on disaster risk assessment and urban disaster reduction measures during this frame. The keywords such as “reduction,” “risk perception,” and “urban resilience” also showed significant intensity, indicating sustained attention to issues related to urban resilience, disaster reduction, and risk perception (Fig. 1.13).

In brief, literature from the WOS database reveals that rapid urbanization has rendered cities highly vulnerable to natural disasters. High population density and

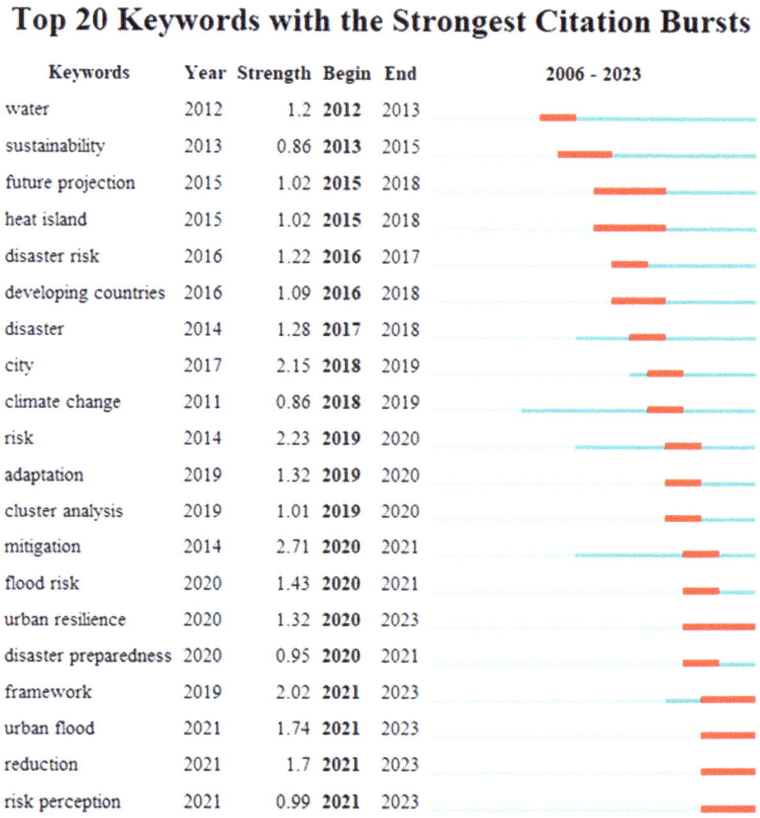

Fig. 1.13 Urban disaster prevention and reduction emergent terms in the WOS database

extensive infrastructure increase potential losses when disasters occur, making urban disaster prevention and reduction crucial for ensuring social stability and resident safety. Research in the database covers a wide range of topics and methodologies, ranging from the effects of climate change to social vulnerability analysis, from technological innovations to the study of complex systems. By conducting a thorough analysis of the vulnerability of different social groups during disasters, more targeted protective measures can be formulated to achieve effective disaster prevention and reduction. Utilizing advanced geographic information systems (GIS), remote sensing technology, data analysis, and artificial intelligence tools can significantly improve the efficiency of disaster warning and response, thus promoting the modernization and intelligentization of disaster prevention and reduction efforts. Future research should further investigate urban disaster prevention and reduction, emphasizing interdisciplinary collaboration and exploring novel theoretical and practical approaches to enhance overall urban resilience.

Current Status of Research on Urban Disaster Prevention and Reduction

With the intensification of global climate change and urbanization, the risk of natural disasters facing cities is steadily rising. Consequently, it is imperative to prioritize disaster prevention and mitigation in urban planning and management. Over the past few years, there has been an increased focus on studying urban disaster prevention and reduction in both academic and practice settings. This has led to significant progress in theoretical advancements and practical insights.

In terms of disaster management planning, Japan is prone to frequent natural disasters, and its citizens exhibit a strong awareness of disaster preparedness. Japan employs a multi-tiered disaster prevention and reduction planning system, encompassing essential disaster prevention planning, operational disaster prevention planning, and regional disaster prevention planning. The "Basic Act on Disaster Countermeasures" guarantees the effective, comprehensive planning and management of urban disaster prevention. Japan's disaster prevention and reduction planning is relatively comprehensive due to the transparency of urban disaster and property information and robust public participation (Jianxin 2009). The United States strengthens its disaster prevention and reduction planning through legislation such as the "Disaster Relief and Emergency Assistance Act" and the "National Emergency Act." Comprehensive planning is oriented towards disaster prevention and reduction. It includes five aspects: the planning process, risk assessment, disaster prevention strategies, coordination of local disaster prevention planning, and planning implementation and monitoring. The planning process is characterized by its iterative and continuous nature (Bin and Lu 2021). China's comprehensive disaster prevention planning can be divided into two categories based on the differences in compilation organizations, procedures, and content: within the urban planning system and outside the urban planning system. Urban Planning mainly focuses on engineering disaster prevention, relying on specific construction projects to implement specialized disaster prevention planning in spatial layouts. The majority of the disaster responses adhere to the relevant provisions of the Ministry of Construction's "Outline of Building Technology Policies." However, they generally lack specificity and wide-ranging applicability. On the other hand, planning outside the urban system is centered around targeted engineering projects aimed at comprehensive disaster prevention and reduction. This approach emphasizes disaster warning, response, emergency handling, and related mobilization and organizational procedures (Lei et al. 2011). Disaster prevention, reduction, and response in the United Kingdom are primarily achieved through legislative means to ensure a complete regulatory system. Government departments cooperate in disaster prevention and relief efforts to ensure clear responsibilities without overstepping authority or omissions. In addition, the United Kingdom encourages broad social participation, ensuring diversity among participating entities. The United Kingdom particular emphasis on enhancing disaster resilience when constructing an emergency management system. The United

Kingdom actively adopts high-tech methods, integrating them into the disaster relief system to improve efficiency and effectiveness (Yanfen et al. 2018).

Therefore, when formulating comprehensive disaster prevention and reduction plans, countries emphasize a holistic approach to construct a thorough and mature planning system. This system encompasses all levels of urban planning, ranging from macro to micro, ensuring a comprehensive approach while addressing technical and engineering challenges. Furthermore, it establishes a robust public participation mechanism. However, rapid economic growth, continuous urban expansion, rising population density, and frequent public emergencies and natural disasters have increased the operational burden on cities, heightening disaster risk and vulnerability. Consequently, there is an increased need for comprehensive urban disaster prevention and reduction planning.

Regarding strategies and methods, various scholars have proposed disaster prevention and reduction strategies from different perspectives, which can be broadly classified into macro- and micro-approaches. At the macro-level, resilience is built nationally to defend against disasters, such as exploring urban construction from a national perspective. Community governance is employed at the micro-level to enhance community resilience, particularly in urban areas where it is more feasible to implement. Consequently, the issue of urban disaster prevention and reduction is studied through the urban resilience perspective. On the macro-level, Ouyang et al. suggest that the government should track the latest operational data of infrastructure during the resistance phase. This will help improve decision-making capabilities, adjust the topological structure of infrastructure during the absorption phase, and establish effective post-disaster communication channels during the recovery phase to enhance urban disaster prevention and reduction capabilities (Ouyang et al. 2012). Zheng Y. asserts that it is essential for all levels of government to reach a consensus on building urban disaster resilience and, more importantly, to focus on enhancing reflective capabilities (Yan 2013). On the micro-level, Carpenter proposes that community adaptability can be indirectly enhanced by supporting social networks through the built environment, thereby improving disaster resilience capabilities (Carpenter 2013). Han D. S. et al. advocate adopting smart technologies to build resilient communities. They argue that the precision and dynamic management of smart technologies can enhance resilient communities' flexibility and adaptability (Donsong et al. 2014).

Therefore, studying urban disaster prevention and reduction strategies from different perspectives broadens the scope of this topic. Timely tracking and adjusting infrastructure data and structures, enhancing decision-making capabilities, and establishing effective post-disaster communication channels can improve overall urban disaster prevention and reduction capabilities. Furthermore, all levels of government must reach a consensus on building disaster resilience, focusing on improving and strengthening reflective and adaptive capabilities. Community adaptability is indirectly improved by supporting social networks through the built environment. The application of smart technologies allows for precise and dynamic management, thereby increasing the flexibility and adaptability of constructing resilient communities. The comprehensive application of these strategies significantly enhances urban disaster prevention and reduction capabilities.

Regarding disaster prevention and reduction mechanisms, most studies are based on the concept of resilient cities. The mechanisms can generally be categorized into three stages. Chen L. et al. proposed a framework for resilient cities that involves three stages in dealing with external disturbance: endurance, resilience, and reconstruction (Li et al. 2017). This framework is based on thoroughly analyzing resilient cities' core concepts and fundamental structure. (1) Endurance Stage: When a city initially encounters external disturbances, a city with strong self-repair capabilities can endure a certain degree of damage without requiring adjustments to the urban system. (2) Resilience Stage: As external disturbances intensify, the urban system continually self-adjusts to adapt to new changes. (3) Reconstruction Stage: When facing greater external disturbances, the city develops new systems that can adapt to the new external conditions, gradually improving economic, environmental, and other aspects within these systems. Zou L. argues that disaster prevention and reduction in resilient cities primarily involve the stages of resistance, recovery, and adaptation (Liang 2018). (1) Resistance Stage: The primary task is to ensure the regular operation of core urban functions, minimizing the negative impacts of disasters. Recovery Stage: The primary task is to restore the damage caused by disasters to the city. (2) Adaptation Stage: The primary task is to optimize the urban system through learning and summarizing experiences to enhance the ability to cope with future risks.

Urban disaster prevention and reduction is an intricate and multifaceted systems engineering project that involves several aspects, such as natural disaster prediction (e.g., early warning, emergency response) and post-disaster recovery (e.g., resistance, recovery, and adaptation). The current research on these mechanisms also follows this sequence. Therefore, future research should explore these mechanisms further by integrating other disciplines and novel technologies.

Problems in the System: Jia J. H. mentioned in his monograph Improving Infrastructure to Safeguard Community Safety that communities, as crucial grassroots social organizations, play a vital role in responding to sudden natural disasters and serve as the frontline for disaster prevention and reduction. Under the guidance of comprehensive disaster reduction demonstration communities, various regions across the country have actively strengthened the construction of community disaster reduction capabilities, considering their specific conditions. This has led to a continuous enhancement of community disaster prevention and reduction capabilities (Junhui 2017). Liu L. F. et al. mentioned in their monograph Disaster Management Systems Providing Support for Disaster Reduction and Relief Decision-Making that a scientific and standardized natural disaster statistics system enhances the timeliness and accuracy of disaster information reporting. Implementing a 24-h disaster emergency duty system improves information communication and monitoring capabilities. A refined and standardized disaster consultation system promotes cooperation among disaster-related departments, facilitating disaster information sharing. A stringent disaster reporting and assessment system promotes executing disaster management responsibilities. As these systems advance and enhance, disaster management levels rise, providing more support for disaster reduction and relief decision-making (Longfei et al. 2017).

Although the concerns raised by the scholars above are significant regarding the construction of disaster prevention and relief systems, it is worth noting that additional problems exist beyond system construction. These include limited interdisciplinary communication, inadequate establishment of comprehensive disaster prevention institutions, slow adaptation of high-tech disaster prevention and reduction measures, and weak public awareness of disaster prevention and reduction practices. These concerns and issues should be tackled in future research and practice.

Review of Research

By reviewing and analyzing the relevant literature, it is evident that the number of research outcomes in urban disaster prevention and reduction has significantly increased. Scholars primarily concentrate their research on disaster planning, disaster prevention, disaster reduction, urban resilience, and urban safety. These studies lay a rich foundation for advancing urban disaster prevention and reduction theory. Despite the advancements made, there is a limited amount of research specifically focused on the application of urban disaster prevention and reduction. The field tends to focus more on theoretical exploration and conceptual framework construction, while practical measures and operational research are relatively less developed. Urban disaster prevention and reduction tasks are of utmost urgency and importance due to natural disasters such as earthquakes, floods, and typhoons and social risks such as increasing population density and aging infrastructure. Hence, research on urban disaster prevention and reduction measures is not merely an academic issue but a significant practical matter concerning urban safety and sustainable development. Although experts and scholars from databases of CNKI and Web of Science have conducted in-depth discussions on research hotspots in urban disaster prevention and reduction from different perspectives, comprehensive and application-oriented research remains insufficient. Despite the progress made in current research, achieving truly effective urban disaster prevention and reduction requires further strengthening of applied research and exploring practical and feasible disaster prevention and reduction measures and strategies based on theoretical research.

In conclusion, future research should prioritize the practical application of urban disaster prevention and reduction. This involves considering the specific circumstances of different cities and leveraging emerging technologies to develop targeted measures and strategies. These efforts aim to enhance cities' disaster prevention and reduction capabilities and resilience, ensuring urban safety and the well-being of residents.

Four Related Concepts in the Field of Disaster Prevention and Reduction Planning

The conceptual framework governing disaster prevention and mitigation needs to evolve from a narrow focus on isolated disaster types to a comprehensive, integrated strategy. More importantly, it should progress from simply diminishing disaster losses to actively mitigating disaster risks. This evolution is crucial for fostering urban environments that are both resilient and secure. The fusion of disaster prevention, risk management, resilience, and safety considerations now stands at the forefront of contemporary research discussions.

First, integrated disaster prevention and mitigation embody a broad spectrum of defensive strategies and actions that cover disaster safety planning, resource allocation, system development, and infrastructure. This approach is characterized by its responsiveness to multiple hazards, engagement of the entire community, and the synergistic application of varied methodologies, orchestrated cohesively across temporal and spatial dimensions.

Second, disaster risk management seeks to mitigate the detrimental effects of disasters through coordinated efforts involving administrative frameworks, technological innovations, and strategic policy enactments. This includes a thorough risk identification, analysis, evaluation, and management process, focusing on the inherent uncertainties associated with these risks and the critical need for proactive measures.

Third, resilience emphasizes the ability of systems to adapt to changes and maintain stability. In disaster response, resilient frameworks are designed to withstand, absorb, adapt to, and recover from adverse events, facilitating rapid transformations. The adaptive cycle, a fundamental principle of resilience theory, elucidates the dynamics of maximizing benefits and minimizing harms through continuous adjustment.

Lastly, urban safety is defined by the capacity to address many security challenges and maintain a baseline of safety. It pursues a vision where every person has an equal right to safety and health, supporting the broader goals of urban development.

Urban Safety

Bibliometric Analysis of Urban Safety

An analysis of 152 CSSCI articles on "urban safety" in the CNKI database from 2000 to 2023 with a one-year time slice reveals a predominant focus on keywords related to "safety." These keywords encompass urban safety, public safety, safe city, disaster prevention and mitigation, and emergency management. The co-occurrence network of urban safety keywords is shown in Fig. 1.14.

Fig. 1.14 Co-occurrence network of urban safety keywords in the CNKI database

According to the time zone view of urban safety keywords, the research on topics related to urban safety commenced earlier in the CNKI database, dating back to 2001, as indicated by the available literature. Initially, the database focused on researching the safety of central cities. Over time, the research direction expanded to encompass public safety, disaster prevention and planning, security planning, security governance, and ecological safety. The CNKI database shows an apparent increase in the prominence of topics of urban safety, smart cities, security prevention, and response, as depicted in Fig. 1.15.

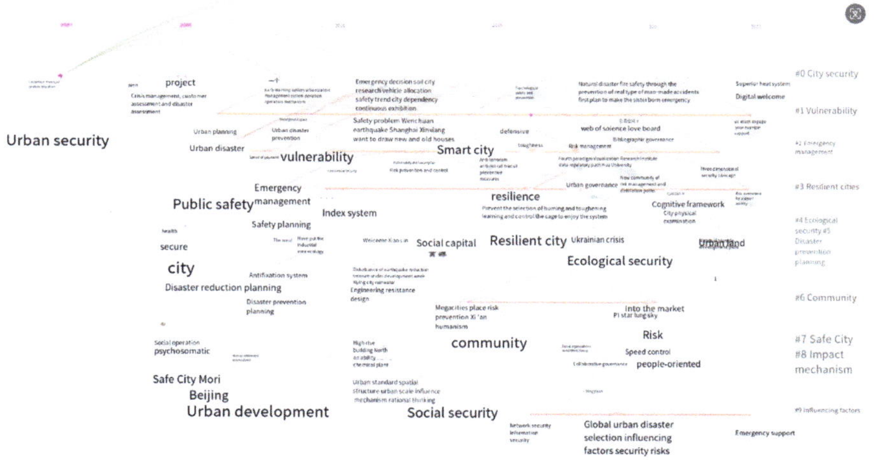

Fig. 1.15 Time zone view of urban safety keywords in the CNKI database

The citation bursts of urban safety keywords reveal that the highest intensity is observed for terms such as "urban safety," "resilient city," "integrated disaster reduction," "risk management," and "urban experience" have the highest intensity. Recent emergent terms include "urban experience," "urban resilience," "resilience governance," "smart city," and "risk management," indicating that these keywords represent trending research topics in the CNKI database in recent years. Furthermore, the research related to urban safety has evolved into topics concerning urban resilience, as depicted in Fig. 1.16.

In summary, the bibliometric study findings reveal a change in urban safety-related literature research in the CNKI database. Initially focused on urban safety, urban disaster, and urban crisis, the research gradually transitioned towards safety planning, disaster prevention planning, urban resilience, and related areas. Research on urban safety problems has been ongoing, starting with the initial identification of the problem and continuing through current discussions on problem-solving and management strategies. Consequently, the research direction concerning urban safety

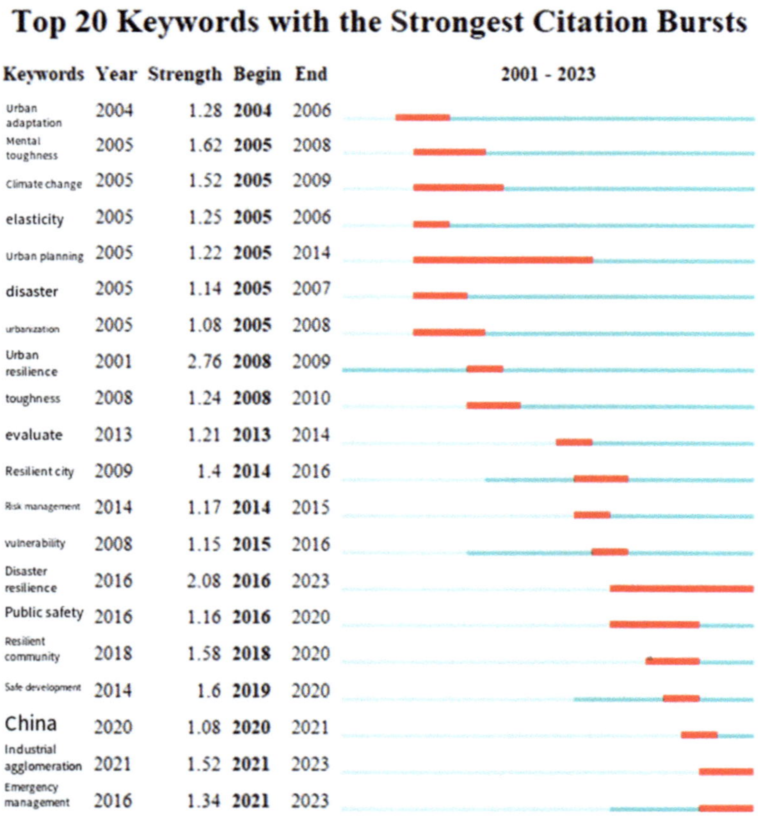

Fig. 1.16 Citation bursts of urban safety keywords in the CNKI database

Fig. 1.17 Co-occurrence map of urban safety keywords in the WOS database

has progressively moved towards safety planning, governance, and urban resilience. Urban safety and resilience research will gain more prominence as it becomes integrated.

An analysis of 144 SSCI articles on "urban security" in the WOS database from 2000 to 2023 with a one-year time slice reveals that research topics on urban security primarily revolve around environmental change, policy, urban crime, public space, climate change, urban surveillance, community safety, and georegulation, as depicted in Fig. 1.17.

Figure 1.18 depicts the time zone view of urban safety in the WOS database. The literature extracts nine themes: "policing," "South Africa," "local hacking," "social geography," "crowd detection," "climate change," "regeneration," "environmental forces," and "local policy." The research on urban safety has been conducted from 2005 onwards, yielding varied degrees and categories of research results. Research on this topic in the WOS database is more abundant and diverse than those in the CNKI database. The research trend regarding urban safety has evolved towards crowd detection, safety governance, and environmental policy direction, which exhibits a more systematic and diversified approach compared to earlier research. Additionally, research related to urban safety resilience has commenced, as depicted in Fig. 1.18.

The most prominent words in terms of emergent word strength are "security," "governance," "policy," "crowd counting," and "space." In terms of the emergence year, "space" and "homeland security" first appeared in 2010, followed by "citizen security," "policy," and so forth. The most recent emergent words are "climate change" and "decision making," indicating that research topics represented by these keywords are currently trending in the WOS database, as depicted in Fig. 1.19.

In summary, research on urban safety in the WOS database primarily focuses on safety response, including policy and population dynamics, among other factors. The

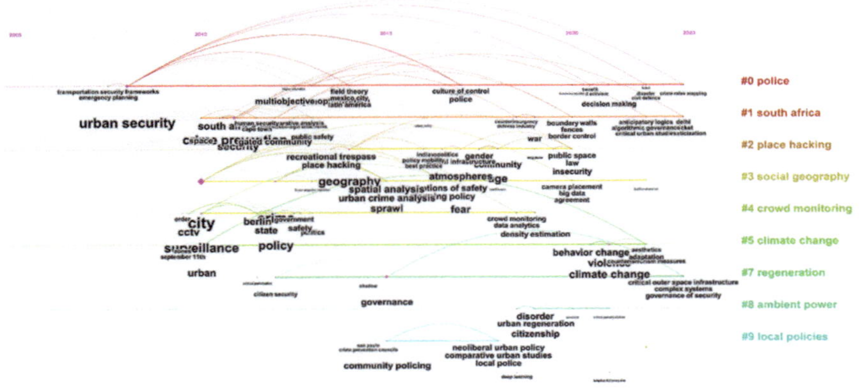

Fig. 1.18 Time zone view of urban safety keywords in the WOS database

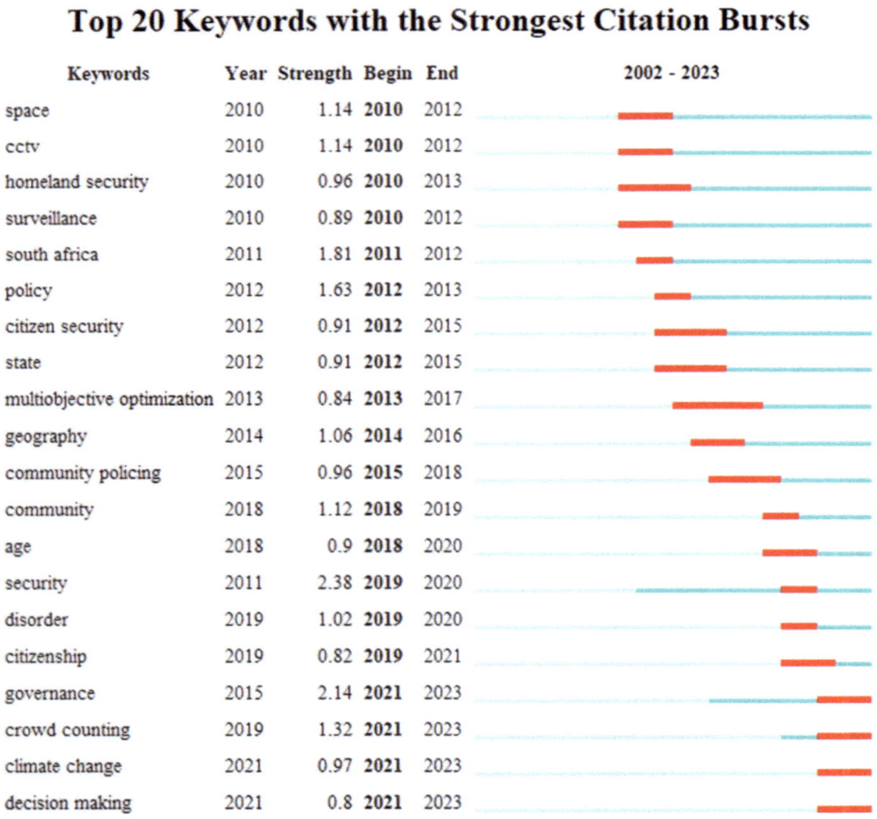

Fig. 1.19 Urban safety emergent words in the WOS database

research tends to be broad, widely disseminated, and reflects rich connotations and characteristics. Recently, research has shifted from focusing on spatial governance and urban network security to examining the impact of the climate environment on urban security and human governance. The advancement of urban security should not solely focus on urban public security research but also integrate current environmental and social factors. Hence, further research is warranted to integrate urban security with urban resilience.

Literature Review on Urban Safety

Urban security is a pivotal issue in contemporary societies. As urbanization, population growth, and advancements in science and technology continue, cities are encountering increasingly complex security challenges. In early urban environments, security primarily pertained to safety issues. However, the modern understanding of urban security has expanded significantly to various factors, including political, economic, social, and environmental dimensions (Yibao 2018).

There has been a surge in research focusing on urban issues, public crises, urban crises, and urban safety challenges. Developments in urban safety can be categorized as follows:

Regarding the connotation of the research analysis: Jin Lei addresses the shortcomings of urban safety management and proposes strategies for building a safe and resilient city, emphasizing the importance of resilience in urban development (Lei 2017). Zhou Jianxin et al. present an evaluation method for assessing urban public safety risks using high-rise buildings as a case study to enhance its applicability in urban safety risk management (Jianxin and Husheng 2013). Zhi Yali identifies the primary obstacles to urban safety management in China as low public participation, limited risk prevention capabilities among the populace, and an inadequate rule of law framework (Yali 2007). Van den Berg et al. examine urban safety and its determinants from three perspectives: regional safety, drug-related issues, and juvenile delinquency. They provide examples from 11 European cities, such as Rotterdam, Antwerp, and Berlin (van den Berg 2006). Ishikawa delves into the historical evolution of open spaces in Tokyo, examining them through urban landscape planning, urban safety, and fire prevention, using Tokyo as a case study. He elucidates the characteristics and main challenges in realizing a safe city (Ishikawa 2002). Wekerle et al. outline the planning principles, design strategies, and implementation approaches for creating safe cities (Wekerle and Whitzman 1995).

By employing connotative analysis, contemporary urban security research delves deeply into the diverse threats and challenges confronting cities, including but not limited to crime, natural disasters, public health crises, and terrorism. This exploration unveils the multidimensional nature of urban security and its underlying factors. Such studies help us understand the reasons and mechanisms behind urban security issues and provide insights and methodologies for formulating targeted and comprehensive solutions. By simultaneously defining the concept of urban safety and conducting thorough examinations of its connotations, these studies offer valuable guidance for city administrators and policymakers. This helps establish and improve mechanisms for urban safety governance, ultimately leading cities towards safer and more livable environments.

Regarding evaluation modeling methods: Huang Ting establishes an evaluation system for assessing the safety status of urban complexes based on four primary indicators: building structure and facilities, safety management, fire safety, and personnel safety. This framework offers insights for enhancing the safety measures in complexes (Ting and Zaipeng 2017). Shan Ning developed an evaluation index system that considers various aspects such as economic and social security, ecological security, public health security, social order security, and network information security for each city. Based on the evaluation outcomes, targeted interventions were proposed (Ning 2017). Jabareen focused on public security from four dimensions: economic, social, institutional, and infrastructure, and developed an urban resilience system for public security accordingly (Jabareen 2013). Luo Yun et al. designed an indicator system aiming for a safe and prosperous society, integrating it with the 12 critical areas of urban safety to evaluate the overall safety status of the city (Yun et al. 2005). Li Ge categorized the factors influencing urban safety. They constructed an urban safety evaluation index system and developed an urban safety evaluation model that integrates the PSR hierarchical analysis model with the BP neural network (Ge et al. 2020).

The evaluation models consider data and indicators from various social, economic, and environmental factors to construct a comprehensive assessment framework that accurately depicts urban safety. The models adapt to different cities' characteristics and requirements, catering to various sizes and developmental stages. It also enhances its practicality and applicability. Most importantly, the evaluation model assesses the current status and trends of urban safety and identifies safety weaknesses and potential risk areas. It provides city managers with a scientific basis for formulating and implementing urban safety governance and mitigation strategies, ultimately leading to safer and more livable environments.

Connotation of Urban Safety

1. Definition of Security and Urban Safety

The concept of security is a dynamic theoretical framework that undergoes continuous enrichment and refinement. The concept of security is initially defined in the Modern Chinese Dictionary as the absence of danger, threats, or accidents. The term "security" denotes a state devoid of danger, threats, or accidents, accompanied by a sense of safety. From a philosophical viewpoint, security is feeling protected inside one's home while feeling unsafe and exposed outside. However, the individual's conception of security is multifaceted. Citizens require a safe, orderly, and stable environment encompassing global peace, social stability, harmony, and personal security. In such an environment, individuals can predict the consequences of their actions, minimizing unforeseen circumstances. In the field of political science, security is contingent upon resource allocation and objective environmental response. Good governance seeks to maintain an ideal and conducive environment, whereas poor governance may precipitate security deterioration, potentially spiraling out of control. Security concerns encompass the safety and well-being of individuals and affairs, constituting a core responsibility of governing authorities and intricately linked to the governance process of the nation.

The concept of security is frequently employed to convey the anticipation and assurance of safety. Certain researchers posit that security is the confidence engendered through human interactions with nature and society. Some scholars argue that security entails individuals or groups extricating themselves from perilous situations, thus alleviating the fear and fostering a sense of safety and comfort. This primarily pertains to averting the horror inflicted by criminals and preserving both physical and mental well-being. Various scholars interpret security differently, leading to multidisciplinary research focusing on different scenarios. The concept's definition and scope vary, with research primarily encompassing disciplines such as sociology and public administration (Peng 2022).

Urban security is defined and conceptualized differently across various disciplines. Disaster science focuses primarily on the prevention and reduction to foster healthy and sustainable urban social systems. Urban security in psychology primarily addresses the psychological safety needs of urban residents. Urban security is considered crucial in sociology for promoting benign social functioning and fostering the coordinated development of society (Xinghua 2023). Conflicts and contradictions arising in emergencies are directly impacted by the effective response and integration within the framework of urban public safety culture. Urban security in economics aims to ensure the coordinated, orderly, and efficient operation of cities, with considerations for future impacts on urban construction and development (Yudong 2021). Overall, urban security involves the effective management of natural disasters and social emergencies, fostering dynamic, balanced, and coordinated development across environmental, societal, and personal health domains. It aims to provide urban residents good order, comfortable living spaces, and personal safety (Xinghua 2023).

Therefore, this book's definition of urban security entails measures encompassing planning, construction, management, and emergency response within a comprehensive urban management framework. The primary goal is to safeguard residents, property, and the environment in the city from various threats, such as crimes, disasters, terrorism, environmental pollution, and public health incidents. It is performed while maintaining social order, ensuring public safety, and promoting sustained and stable urban growth.

2. Characteristics of Urban Security

Compared to other types of security, urban security possesses unique attributes.

(1) Complexity and Diversity of Influencing Factors

Urban safety is a multifaceted concept that includes multiple fields and dimensions. It encompasses not only natural disasters but also man-made disasters, public safety, environmental safety, and other aspects. The intricate and varied influencing factors manifest across all urban planning, construction, and management facets. It is imperative to comprehensively evaluate the impact of diverse safety factors and implement corresponding measures for prevention and management.

Urban safety's complex and diverse nature necessitates cross-sectoral and interdisciplinary cooperation involving governments, enterprises, communities, and residents in urban safety management. During urban planning, safety requirements related to transportation, environment, and disaster prevention must be considered, and corresponding plans and policies formulated. In urban construction, safety design and measures such as safe construction are imperative to ensure the safety of buildings and infrastructure. Besides, effective urban management demands the establishment of robust safety monitoring, early warning, and emergency response mechanisms to address various safety risks promptly. City managers must embrace a holistic approach, considering specific safety domains, the interaction between all aspects, and their collective benefits. Only through a comprehensive approach can the overall safety level of the city be effectively enhanced, ensuring the safety and well-being of urban residents.

(2) Uncertainty of Influencing Factors

The uncertainty of influencing factors is a significant characteristic encountered in urban safety. Urban safety management involves dealing with the uncertain effects of different factors, such as natural disasters, human actions, and the social environment. This uncertainty challenges city managers, making it challenging to predict and plan a secure trajectory for cities. The timing, location, and intensity of natural disasters (e.g., earthquakes, floods, and typhoons) are frequently difficult to predict accurately, hindering city managers' ability to take proactive measures to mitigate disaster-induced damage. Human factors, such as terrorist attacks, traffic accidents, and criminal activities, often occur suddenly and unpredictably, posing significant challenges to urban security. City administrators find it arduous to anticipate and preemptively address the security risks posed by human factors, necessitating readiness to respond promptly. Moreover, uncertainty in the social environment impacts urban safety. The complexities and uncertainties of factors such as socio-economic development, population mobility, and the political landscape often create challenges for urban security.

(3) Degree of Safety

The degree of security is a pivotal aspect of urban safety, reflecting a city's level of security and its capacity to cope with diverse potential threats and risks. The degree of security plays a crucial role in determining the overall safety of a city. Natural disasters' frequency, intensity, and scope directly influence a city's security level.

Cities located in high-risk areas or lacking effective disaster prevention and reduction measures will experience more significant impacts on their safety levels. The security level of a city is directly reflected in its performance in terms of crime rate, social order, and public safety. A low crime rate and favorable public order environment enhance the city's safety level and residents' sense of security. The quality of infrastructure, including transportation, water supply, power supply, and communication, directly influences a city's emergency response capability and resilience to disasters. Cities equipped with robust infrastructure systems can more effectively address diverse security threats. The level and efficacy of government governance significantly impact the effectiveness of security management in cities. Efficient government governance facilitates the provision of quality public services, market order regulation, and social stability maintenance, ultimately enhancing the degree of security. Urban safety is not merely a singular metric but a culmination of various factors. Improving urban safety requires collaborative efforts by government, enterprises, communities, and residents. This involves strengthening disaster prevention and reduction measures, enhancing the social security environment, upgrading infrastructure construction, bolstering government governance capacity, and fostering safety awareness among residents. These measures collectively enhance the degree of safety, guaranteeing the safety of residents' lives, property, and the environment.

(4) Evaluability of Safety

Another crucial aspect of urban safety is its evaluability, which denotes the degree to which the level of urban safety can be quantified, assessed, and compared using specific criteria and indicators. To comprehend the current safety situation, implement appropriate measures, and enhance the safety level of the city for city managers and stakeholders, it is crucial to have an effective measuring method to assess the safety levels. Urban safety hazards can be divided into two main categories: natural disasters and man-made disasters. Natural disasters caused by natural phenomena can be further classified into geological, climatic, and compound disasters. Man-made disasters are mainly caused by people's negligence, improper management, old equipment, electrical short circuits, and other reasons. These two categories of factors can be further decomposed to establish an indicator system for urban safety, which indicates that urban safety is still evaluable. Based on specific criteria, we can ascertain which cities are safe and which are not through people's sensory perception. The criteria for identifying a secure city are not fixed and evolve in response to changing circumstances. The perceived security threat in the same city or region may vary over time (Fengliang 2012).

Urban Resilience

Bibliometric Analysis of Urban Resilience

Keywords serve as concise and generalized summaries of the article's subject matter. They illuminate research trends, intrinsic associations, and the level of significance within the field of urban resilience based on criteria such as relevance, frequency, and emergence. Analyzing keyword frequency and clustering in academic literature aids in intuitively understanding the research's primary concepts and critical insights, as depicted in Fig. 1.20.

An analysis of 380 CSSCI articles on "urban resilience" in the CNKI database from 2000 to 2023 with a one-year time slice reveals notable contributions from terms such as "urban resilience," "resilient city," "resilience," "economic resilience," "disaster resilience," "evaluation," and "resilience governance," among others. Figure 1.21 depicts the sequence diagram of keyword clustering results, organizing each clustering category from left to right based on the chronological order of documents. Over the longitudinal timeline, the evolving trends in research hotspots are related to urban resilience content (Linjun et al. 2023).

The analysis of the chronological chart reveals that domestic research on urban resilience content commenced as early as 2012, aligning with the developmental node

Fig. 1.20 Co-occurrence network of urban resilience keywords in the CNKI database

Fig. 1.21 Time zone view of urban resilience keywords in the CNKI database

of resilience concepts applied to cities from other disciplines. The literature yielded ten distinct themes, including “urban resilience,” “resilient city,” “resilience,” “economic resilience,” “resilient governance,” “resilience evaluation,” “urbanization,” “digital economy,” “disasters,” and “urban agglomerations.” The research trajectory spans from the initial exploration of relevant concepts to urban resilience and various resilience-related topics, gradually encompassing urban disasters. The role of the digital economy in urban resilience and the dynamics of urban agglomerations are illustrated in Fig. 1.21.

The highest-frequency keywords reflect the influential research domains during a specific timeframe, signifying topics that have garnered considerable attention in the academic community and indicating the development trends of cutting-edge research areas (Wenya et al. 2023). The highest-frequency keywords include “climate change,” “resilience,” “urban planning,” “emergency management,” “industrial agglomeration,” “urbanization,” and others. Earlier keywords include “urban adaptation,” “mental resilience,” and “urbanization.” Conversely, keywords that appear later include “safe development,” “industrial agglomeration,” and “emergency management.” This indicates that these keywords have recently become prominent trends in China’s academic landscape. Urban resilience-related research has evolved progressively from initial themes of urbanization and urban adaptation to encompass safety development, emergency management, and other coping-related topics, as depicted in Fig. 1.22.

An analysis of 219 SSCI articles on “urban resilience” in the WOS database from 2000 to 2023 with a one-year time slice reveals notable contributions from terms such as “adaptive capacity,” “adaptation,” “community resilience,” “framework,” “indicators,” and “ecological resilience.” Urban resilience research predominantly

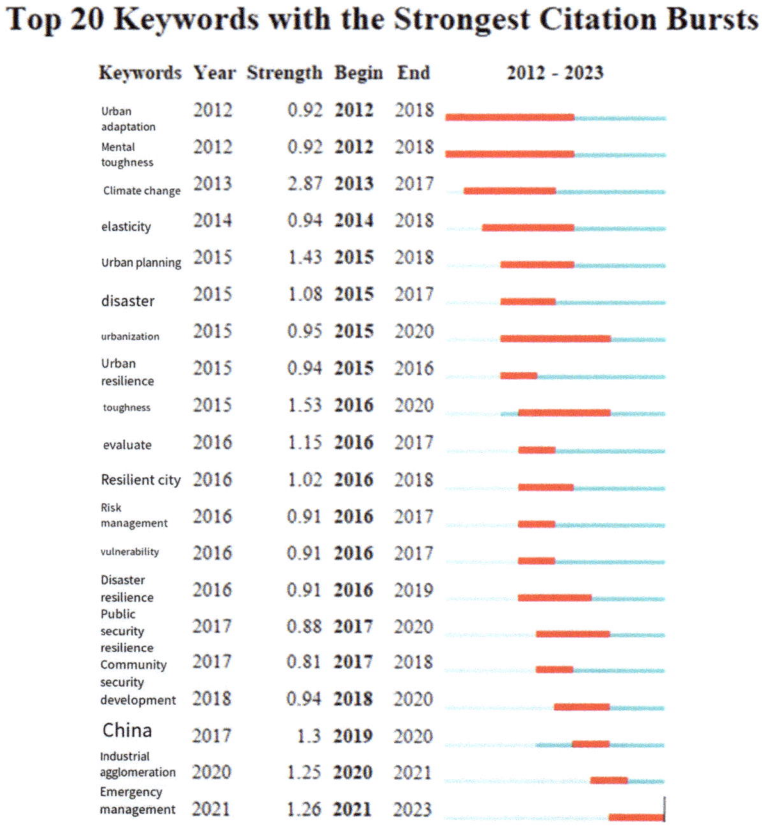

Fig. 1.22 Highest-frequency of urban resilience keywords in the CNKI database

centers on the diverse capabilities of urban resilience alongside relevant indicators and frameworks, as depicted in Fig. 1.23.

Examination of the temporal distribution reveals that research on urban resilience in the WOS database dates back to 2009, showing a slight disparity from that in the CNKI database. The WOS database primarily emphasizes governance, disaster response, resilience evaluation, and governance capacity. However, recent studies have predominantly focused on resilience modeling and various hazard types, as illustrated in Fig. 1.24.

The citation bursts keywords reveal the top occurrences to be "ecology," "climate change adaptation," "urbanization," "climate change," and "adaptation." Earlier occurrences of keywords include "adaptation," "challenges," "biodiversity," "ecology," "climate change adaptation," and "urban planning." Conversely, the highest-frequency keywords appearing later in the year encompass "urbanization," "ecosystem services," "climate," and "hazards." This indicates that recent trends in urban resilience research in the WOS database have progressively shifted towards

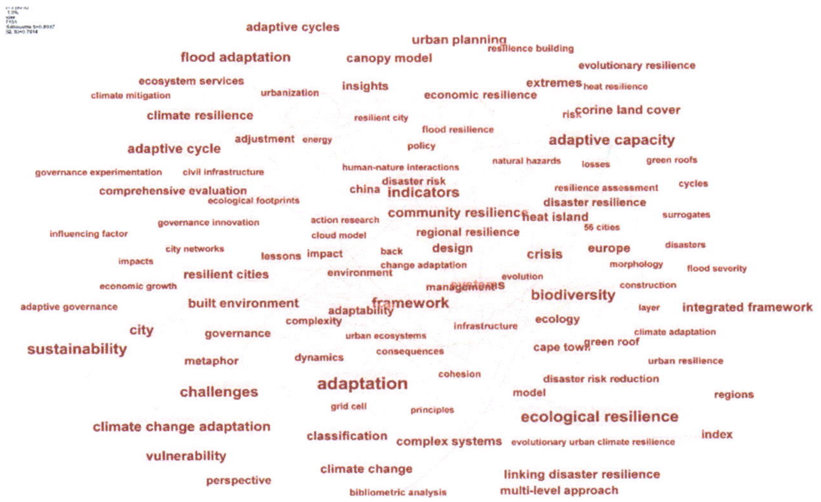

Fig. 1.23 Co-occurrence network of urban resilience keywords in the WOS database

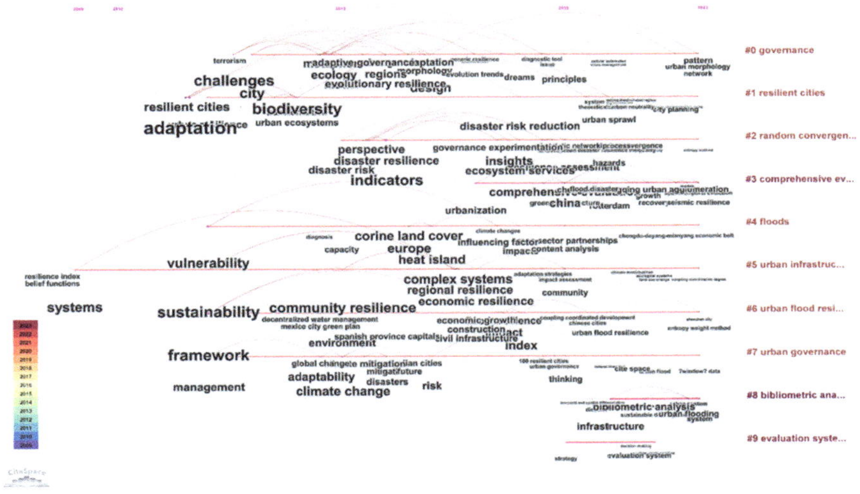

Fig. 1.24 Time zone view of urban resilience keywords in the WOS database

examining cities confronting climate change, biodiversity, and various types of disasters, as depicted in Fig. 1.25.

In summary, analysis of the two databases reveals that initially, studies on urban resilience shared a focus on responding to climate change, disasters, and related factors. Over recent years, studies in the CNKI database have increasingly delved

Top 20 Keywords with the Strongest Citation Bursts

Keywords	Year	Strength	Begin	End
adaptation	2012	1.83	**2012**	2016
challenges	2013	1.32	**2013**	2020
biodiversity	2014	1.25	**2014**	2015
ecology	2015	2.69	**2015**	2019
climate change adaptation	2015	2.52	**2015**	2020
urban planning	2015	1.66	**2015**	2018
environment	2015	1.22	**2015**	2016
lessons	2015	1.18	**2015**	2017
adaptability	2015	1.14	**2015**	2016
policy	2016	1.58	**2016**	2020
regions	2016	1.18	**2016**	2020
perspective	2016	1.17	**2016**	2019
urbanization	2018	2.02	**2018**	2019
climate change	2015	1.98	**2018**	2021
ecosystem services	2019	1.56	**2019**	2020
metaphor	2015	1.42	**2020**	2021
climate	2020	1.31	**2020**	2021
model	2015	1.47	**2021**	2023
rotterdam	2021	1.41	**2021**	2023
hazards	2021	1.41	**2021**	2023

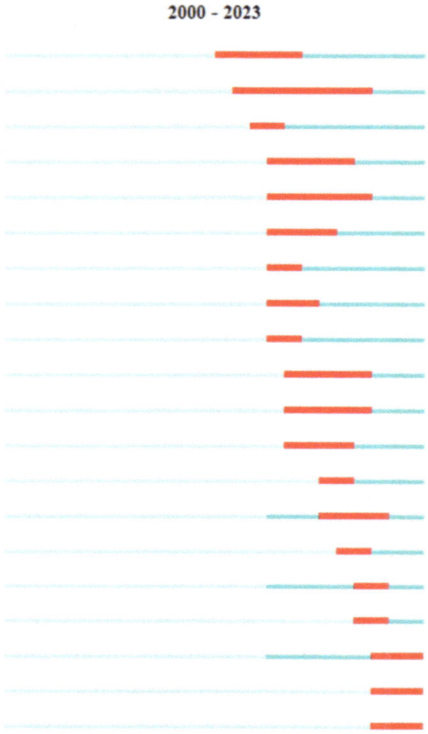

Fig. 1.25 Citation bursts of urban resilience keywords in the WOS database

into security management, industrial agglomeration, and emergency management. In contrast, research on urban resilience in the WOS database has shifted towards focusing on resilience modeling and disaster response. In conclusion, the discourse on urban resilience spans across various databases.

Literature Review on Urban Resilience

1. General Development of Urban Resilience

Urban resilience has emerged as a crucial concept in urban planning and its associated disciplines. The focus of academics and policymakers on urban resilience theory has steadily increased. Research on urban resilience has evolved from theoretical analysis to a deeper exploration of internal mechanisms. The comprehension of resilience in urban environments and its implications for national development and construction has become increasingly profound. Notably, initiatives like the "sponge city" and the "park city" serve as prime examples of this growing awareness (Xinyu and Qiao 2019). Urban resilience can be divided into two distinct stages from a comprehensive perspective. From 2012 to 2017, the first stage was characterized by gradual progress. Urban resilience research evolved with a specific focus on the conceptual framework, characteristics, and evaluation criteria of urban resilience. Research endeavors included exploring the application of resilience theory to mitigate climate change-induced disasters, delineating the distinction and relationship between elasticity and resilience, and devising strategies within urban planning to address the adverse impacts of urbanization. Subsequently, rapid development ensued from 2018 to 2021, building upon the groundwork laid in the preceding years. Research efforts intensified with a focus on expanding and deepening study topics to enhance urban resilience. Emphasis was placed on the adoption of nature-based solutions and the establishment of urban ecosystems. The research also explored macro-morphology, examining the relationship between resilience and urban form. As a result, strategies were devised for urban building and energy refurbishment, fostering the development of smart cities and integrating sustainable practices and urban services. Information and communication technologies (ICTs) were also utilized to enhance urban spaces, while urban agriculture was promoted to bolster cities' food resilience capacity (Yujie and Xunqing 2023). Nevertheless, variations in research perspectives lead scholars to emphasize different aspects of urban resilience. For instance, scholars in the natural sciences prioritize the physical aspects of urban systems, whereas those in the humanities and social sciences focus more on socio-economic dimensions. Disciplinary disparities partially restrict the advancement and implementation of urban resilience (Ying and Wangfeng 2022). Several essential features can characterize urban resilience research. The theoretical framework has transitioned from a single-system to a multi-system complex framework, resulting in broader coverage and evident disciplinary intersections. Urban resilience research has delved deeper beyond initial engineering investigations into disaster prevention and control, urban culture, and other aspects. There is an increased focus on resilience research at the level of social governance, reflecting broadened research perspectives (Zezhao and Runnan 2023). Researchers focusing on urban resilience have proposed multi-level enhancement programs ranging from local governments to urban systems, encompassing macro- to micro-levels. Furthermore, emphasis has been placed on enhancing strategies across multiple fields, including urban planning, social governance, and government emergency management. Such comprehensive research and exploration can greatly enhance cities' resilience, enabling them to adapt better and respond to diverse challenges and risks.

2. Current status of Urban Resilience Development

Urban resilience can be classified into the following categories:

Theoretical frameworks of urban resilience: Jha and colleagues examined the urban resilience framework across four dimensions: infrastructure, institutional, economic, and social resilience (Jha et al. 2013). Pickett and co-authors elaborated on urban resilience within the Human Ecosystem Framework, drawing on ecological, socio-economic, and planning disciplines (Pickett et al. 2004). Desouza developed a framework for urban resilience, focusing on factors such as culture and process dynamics within the city (Desouza and Flanery 2013). Lorenzo Cheller proposed that the theoretical framework for urban resilience should revolve around three dimensions: resilience (system maintenance), transformation (gradual change of the system), and transformation (repositioning of the system) (Chelleri 2012). Zhao Ruidong and colleagues summarized three perspective frameworks: disasters, urban governance, and complex adaptive systems. They discussed the essential viewpoints, framework significance, and applications, advocating for the gradual establishment of a unified and recognized theoretical framework (Ruidong et al. 2020). Anna Bozza presented a new framework for developing urban resilience, aiming to quantify the resilience of urban systems while ensuring sustainability (Bozza et al. 2015). The United States Agency for International Development (USAID) introduced a theoretical framework for urban resilience in developing countries (Frankenberger and Nelson 2013). The action research for community-based adaptation (ARCBA) focused on five major residential areas, approaching them from five perspectives: evaluation systems, resilient government decision-making systems, business engagement, and academic institutions of knowledge, information, and innovation (Ayers et al. 2012).

The theoretical research framework for urban resilience has progressively evolved, becoming profound and more diversified. Initially, research on urban resilience mainly concentrated on cities' ability to respond to natural disasters, emphasizing establishing effective disaster prevention measures. Over time, this research has expanded to encompass various dimensions, transitioning from static to dynamic evolution and integrating multidisciplinary approaches. Urban resilience assessment strongly focuses on participatory research to ensure a comprehensive, sustained, and practical assessment. Undoubtedly, these practices and studies contribute to the advancement of theoretical frameworks.

The evaluation methodologies of urban resilience: The field of urban resilience assessment still lacks a universally recognized evaluation system or standard. This is mainly because of its wide-ranging scope and diverse research topics, which remain among the significant challenges in urban resilience research. The specific content of urban resilience evaluation comprises two main aspects: evaluation methods and evaluation indicators. (1) For evaluation methods: Jonas et al. selected 25 indicators from five aspects to establish an evaluation system for urban resilience (Joerin et al. 2014). Hudec et al. selected 12 indicators to evaluate and analyze the differences in resilience levels among various cities, focusing on social, urban economic, and urban community management aspects (Hudec et al. 2018). Orencio constructed a resilience indicator system for coastal cities to address flood disasters, selecting 12 indicators covering seven aspects and analyzing resilience levels using hierarchical analysis (Orencio and Fujii 2013). Shi Xixi et al. assessed the ecological security status of various cities using 35 indicators from social, economic, and natural perspectives (Xixi and Li 2020). Zhu Min developed a comprehensive evaluation index system of sponge city resilience, employing the fuzzy Delphi method and comprising five dimensions and 25 evaluation indicators (Min et al. 2023). (2) For evaluation methods: Lam et al. utilized the rim model to evaluate the resilience of

25 Caribbean countries in response to hurricanes (Lam et al. 2015). Zhang Huiming et al. employed the entropy weights TOPSIS method to construct a comprehensive evaluation model, enhancing the evaluation system for a city's flood resilience across society, environment, economy, and management aspects (Zhang et al. 2021a). Zhao Li performed a thorough evaluation and comparative analysis of spatial and temporal characteristics of resilience in 13 cities. It was accomplished using the entropy-weighted and natural cross-sectional methods (Li et al. 2023).

Consequently, the evaluation methodologies have shifted from relying on quantitative indicators and statistical data to focusing on comprehensive and multidimensional assessment. This involves integrating quantitative and qualitative assessments, as well as the utilization of artificial intelligence and big data for evaluation purposes. Evaluation content primarily adopts a multidimensional approach, considering factors including social resilience, economic resilience, environmental resilience, and governance resilience. The assessment emphasizes comprehensiveness, dynamism, participation, and operability to guarantee that the evaluation process and outcomes are successful, practical, and sustainable. This aims to support cities' sustainable development and address diverse challenges.

Urban Resilience Enhancement Programs: Previous risk management strategies primarily focused on bolstering defenses to enhance cities' ability to respond to risks. However, this approach often makes cities maladaptive and ineffective in responding to sudden onset risks. In contrast, urban resilience emphasizes the capacity of urban systems to adapt, respond, recover, and learn from risks. This implies that in the face of severe disasters and crises, urban systems can maintain stability, withstand shocks, learn from them, and grow stronger. Zhang Qin et al. proposed measures to enhance the modernization of urban community resilience governance across three dimensions: path, structure, and capacity. They also outlined a realization path to improve the effectiveness of urban community resilience governance, addressing issues such as community emergency response capacity, social capital dispersion, lack of comprehensive planning for community risk prevention and control, fragmented governance resources, and information clutter (Qin and Jiacheng 2023). Jack proposed the concept of resilience enhancement through multifunctionality, redundancy, modularity, diversity, multi-scale networking, and adaptive planning and design (Ahern 2011). Florinda developed a framework of urban resilience encompassing four dimensions: digital city, smart city, sustainable city, and improved quality of life (Matos et al. 2017). Bruzzone argued that enhancing urban resilience is influenced by long-term strategies, investment capacity, and engagement from citizens, businesses, and the third sector (Bruzzone et al. 2021).

Consequently, a significant portion of the current research concentrates on the role of governmental leadership in urban governance while neglecting the pathways through which enterprises, social organizations, and the public can effectively participate. The collaboration between these stakeholders is crucial to urban governance's methods and efficacy. This collaboration holds significant theoretical importance for understanding resilient governance.

Resilience and Urban Resilience-Related Research

1. The Concept of Resilience

The term resilience originates from the Latin word "resilio," meaning "to return to the original state after damage." It has become an emerging concept in research across various disciplines. The term "ecosystem resilience" was initially introduced by American scholar C. S. Holling in 1973. He defined it as the persistence of natural systems in response to ecosystem changes resulting from natural or anthropogenic factors (Holling 1973). Subsequently, other scholars have provided different definitions of "resilience." Carpenter explored the concept of resilience in social systems. He defines resilience as the system's ability to endure perturbations, a vital quality for societies facing developmental changes. The system must preserve society's functionality, structure, self-organization, and capacity for learning to maintain the attribute (Carpenter et al. 2001). Mileti viewed resilience as a system's ability to withstand severe natural events without significant productivity, quality of life losses, and external assistance (Mileti 1999). Adger defined resilience as the capacity of human communities to sustain social infrastructure amidst changes in the external social and political environment, emphasizing recovery time from stress and access to crucial resources like land and finances (Adger et al. 2005). Bruneau defined resilience as mitigating disasters by minimizing their impact, conducting recovery operations to reduce societal damage, and addressing future earthquake repercussions (Bruneau et al. 2003). Meerow defined resilience in systemic and adaptive terms, describing it as the capability of an urban system and its socio-ecological and socio-technical networks to sustain or swiftly restore desired functions amidst disturbances while rapidly adapting to present and future changes (Meerow et al. 2016). The International Intergovernmental Panel on Climate Change (IPCC) regarded resilience as the capacity of a system to absorb disturbances while preserving its fundamental structure and functioning, thereby enabling self-organization and adaptation to stress and change (Change 2007). Ahern contended that resilience underscores a shift from security defense to security without apprehension, as systems possess the capability to reorganize or recover amidst change or disturbance without altering their fundamental functions and components (Ahern 2011). Gordon defined resilience as the capacity of an engineered facility or natural system to withstand, absorb, and transfer kinetic energy while preserving its structural integrity (Gordon 2009).

Indeed, the term "resilience" has been employed in diverse contexts over the past 50 years since its inception (Table 1.2).

In general, this section defines "toughness" as the capacity of a system or material to preserve its structural integrity and functionality when exposed to external impacts or stresses.

2. The Process of Resilience Development

Over time, the concept of resilience has gradually permeated various disciplinary fields, including the natural and social sciences, leading to its evolution and application across diverse domains. This evolution has led to a deeper understanding of

Table 1.2 Definition of the concept of resilience based on the research field

Areas of application	Relevant definitions
Economics	Resilience refers to cultivating a diversified economic structure, enabling the subject of study to adapt and navigate economic turbulence effectively
Sociology	Resilience refers to the capacity of social groups to manage and adapt to social change through self-organization within the society
Disaster science	Resilience refers to the capacity of the subject under study to withstand natural hazards, diminish their impacts, and restore the system to a stable state
Urban design field	Resilience primarily entails cities proactively adapting to environmental changes and climate change through strategic infrastructure layout, land use planning, establishment of blue-green networks, development of transportation systems, and other measures aimed at bolstering the resilience of urban systems (Wei 2023)
Engineering	Toughness refers to the ability of a metal to absorb energy and deform plastically without fracturing when subjected to an external force and to return to its original state after the force is removed (Yiwei 2023)
Physics	Toughness is used to describe and measure a material's ability to resist compression and the maximum pressure it can withstand (Reghezza and Rufat 2015)
Supply chain management area	"Resilience" in the context of supply chains refers to the capacity of the internal structure and control mechanisms of the supply chain to promptly respond and recover from disruptions, thereby ensuring the continued operation of the supply chain despite unforeseen events (Ponomarov and Holcomb 2009)

the adaptability and robustness exhibited by systems or individuals when confronted with challenges and stresses. Essentially, resilience has continued to evolve across various disciplines, offering novel perspectives and approaches for comprehending and addressing complex challenges. However, resilience has undergone three theoretical innovations: engineering resilience, ecological resilience, and socio-ecological resilience (Yiwen and Jiang 2015).

(1) Engineering Resilience

In the 1980s, the prevailing notion of toughness revolved around responsiveness, particularly in engineering. Engineering resilience denotes the capacity of an object to revert to its original state following deformation caused by external forces. From an engineering perspective, the system has a distinctive equilibrium state with unlimited resilience potential to recover from disturbances. This means that all external disruptions can be predicted, allowing the system to quickly return to its original balanced state due to its natural self-organizing abilities. Indeed, this robustness embodies a form of resilience, emphasizing efficiency, stability, and predictability. The resilience of a system is gauged by the speed or timing of its recovery. Folke argued that a system can return to its pre-perturbation equilibrium by resisting, mitigating, and absorbing perturbations (Folke et al. 2002). Similarly, Berkes and Folke contended that engineering resilience underscores a system's capacity to sustain stability or rapidly reestablish equilibrium. This can be assessed by its resistance to external perturbations or the time it takes to reach equilibrium after a disturbance (Berkes and Folke 1994). Engineering resilience represents the capability of an engineering system to uphold its functionality, integrity, and stability amidst diverse external challenges, uncertainties, and unforeseen circumstances. The efficacy of this "engineering toughness" hinges on the duration and velocity of recovery following system impact; the shorter the recovery time, the greater the resilience of the system.

(2) Ecological Resilience

Ecological resilience expands the concept of engineering resilience, but it possesses particular distinctions in its characteristics. The main difference is that engineering resilience assumes a system possesses only one stable state; however, ecological resilience acknowledges the potential existence of multiple equilibrium states within a system, which can transition into other stable states. Gunderson employed the ball-and-cup model to more intuitively illustrate the distinction between the two concepts of resilience (Gunderson 2003). Ecological resilience is frequently employed to describe the persistence or adaptability of natural systems amidst alterations in external environmental factors and internal human factors. This concept comprises two main definitions (Pimm 1984). The initial definition concentrates on the stability of an ecosystem near a specific equilibrium state, measured by the system's resistance to perturbation and its recovery speed to return to the original equilibrium state following disturbance. The subsequent definition emphasizes the instability of ecosystems and centers on assessing the conditions necessary to induce a transition from one state to another. This is quantified by the magnitude of perturbation that precipitates an abrupt shift in the system's equilibrium. The research framework for resilience in ecology encompasses four dimensions: latitude (i.e., the range of system resilience from equilibrium), resistance (i.e., the system's resistance to external perturbations in maintaining equilibrium), precariousness (i.e., the proximity of the system's current state to the critical collapse threshold), and relatedness (i.e., the hierarchical interrelationship of components within the perturbed system) (Gunderson 2000).

(3) Socio-ecological Resilience

Ecological resilience expands upon resilience theory and the dynamics of natural ecosystems by integrating perspectives and attributes from engineering resilience. As systems research progresses, scholars have recognized that ecosystems are simultaneously influenced by social and economic behaviors across multiple scales, resulting in continuous flux. Pickett introduced the concept of socio-ecological resilience, which includes providing natural capital for human development, enduring the impacts of environmental change and human activities, and maintaining a balance between human needs and natural carrying capacity in the short and long term (Pickett et al. 2004). Socio-ecological resilience depends on the capacity of integrated social and ecological systems to be self-sustaining and adaptable in response to change, supported by a blend of social and ecological mechanisms. In particular, socio-ecological resilience encompasses the system's capacity to absorb and adapt to perturbations, alongside its ability for renewal and transformative development (Folke 2006). As the social-ecological system integrates closely linked social and ecological dimensions with mutual constraints and feedback loops, achieving resilience necessitates responses from both dimensions. Focusing exclusively on the adaptive and organizational capabilities of the social aspect while regarding the ecological dimension as a "black box" can lead to systemic pitfalls. Resilience achieved through this approach typically comes at the expense of ecological adaptive capacity. Similarly, focusing exclusively on the ecological dimension can lead to narrow and misguided decision-making (Smit and Wandel 2006). Socio-ecological resilience places a high importance on the processes of renewal, repair, and self-organization over returning to equilibrium. Therefore, resilience represents a capacity for change, signaling a positive move towards sustainability. The perturbations associated with this approach are not always detrimental but frequently serve as catalysts for system optimization, innovation, and sustainable development (Walker et al. 2009).

Socio-ecological resilience pertains to the capacity of intricate socio-ecological systems and their subsystems to adapt and transform in the face of disturbances, shocks, and changes. Unlike conventional frameworks for change management, this definition prioritizes the system's long-term dynamic evolution and developmental capacity, making it particularly appropriate for studying the resilience inherent in cities (Table 1.3).

3. Key Dimensions of Resilience

Although the concept and connotation of resilience have not yet been fully harmonized across various disciplines and institutions, some of its fundamental attributes and critical dimensions have been extensively debated. Clarifying the multidimensional characteristics of resilience serves as a "bridge" to deeper comprehension of its connotation and further measurement of resilience. Engineering resilience is characterized by its capacity to endure stress, as demonstrated by robustness, resistance, stability, or durability. Ecological resilience views resilience not merely as a single capability but as a synthesis of multiple capabilities, expanding the scope of resilience dimensions. Under the influence of social sciences, socio-ecological resilience has been construed as an emerging attribute encompassing two additional dimensions: adaptive capacity, denoting the capability to assimilate experience and knowledge, modulate responses to evolving external and internal stimuli, and sustain functionality; and transformational capacity, defined as the ability to engender an entirely new system when prevailing ecological, economic, or social structures render the existing system unsustainable, thereby creating a novel paradigm. The four resilience capacities, namely resistance, absorptive, adaptive, and transformational, are considered the latest and most widely acknowledged fundamental attributes and universal dimensions of resilience (Ge et al. 2016). These four attributes encapsulate the fundamental principles of engineering, ecological, and socio-ecological resilience (Hongrong 2023).

Table 1.3 Comparison of three resilience types from different evolving season

Toughness types	Definition of concepts	Points of interest	Characteristics	Objective	Theoretical foundation
Engineering resilience	The speed at which the initial velocity can be restored after a disturbance	Unique homeostasis	Ordered and linear	Restore unique homeostasis	Engineering theory
Ecological resilience	Stability and magnitude of perturbations that ecosystems can absorb	Multiple steady states	Complex and nonlinear	Homeostasis change	Ecological theory
Socio-ecological resilience	The capacity of complex systems to mitigate, adapt, and learn from risks following perturbations	No steady state, dynamic cycling across scales	Adaptive cycles, dynamic interactions	Adaptation, self-organization, learning and innovation	Theory of complex adaptive systems and theory of evolutionary economic geography

4. The Concept of Urban Resilience

The concept of resilience has emerged with the continuous evolution of time and the ever-changing environmental landscape. It closely connects complex systemic characteristics inherent to cities, thereby spawning the concept of urban resilience. The concept of a "resilient city" was first introduced at the United States Ecology annual meeting in 2002. Urban resilience entails varied definitions stemming from divergent research methodologies, objectives, and fields of study adopted by different countries, regions, and scholars. Presently, resilience research primarily concentrates on addressing specific threats (e.g., climate change, flooding) and encompassing all inherent risks within cities, with urban resilience serving as an aspirational objective and the concept of resilience broadened as a cognitive framework (Folke et al. 2010). The concept of "resilience" has prompted more academic investigation into "urban resilience."

This section elucidates the conceptual definition of urban resilience by analyzing highly cited literature from domestic and international sources and studies conducted by international organizations (Table 1.4).

Through an analysis of definitions provided by scholars and organizations, it is evident that urban resilience arises from the acknowledgment of threats to city development, emphasizing the city's capacity to respond, adapt, and transform amidst crises such as natural disasters, climate change, and political instability. Another

Table 1.4 Conceptualization of urban resilience by different scholars

Authors	Research areas	Definition
Godschalk (2003)	Engineering	Urban resilience refers to a sustainable network comprising both urban physical systems and human communities
Alberti et al. (2003)	Biological sciences/ environmental sciences	Urban resilience is defined as the capacity of a city to withstand change before undergoing structural and procedural reorganizations
Pickett et al. (2004)	Biological sciences/ environmental sciences	Urban resilience refers to the ability of an urban system to adapt to changing circumstances
Rockefeller Foundation (2014)	Geography	Urban resilience refers to the capacity of individuals, communities, institutions, businesses, etc., to survive, adapt, and undergo further development or transformation in response to risks and shocks
Jha et al. (2013)	Science of disasters	Urban resilience consists of infrastructure, institutional, economic, and social resilience dimensions, with infrastructure resilience being pivotal, as it influences the seamless functioning of critical infrastructure and the city's capacity for emergency response
Tyler and Moench (2012)	Social sciences/ environmental sciences	Urban resilience encourages the exploration of innovation and adaptation to assist cities in recovering from stresses and shocks, whether predictable or unforeseen
Leichenko (2011)	Social sciences/ environmental sciences	Urban resilience entails the capacity to withstand diverse shocks and stresses
Ahern (2011)	Environmental sciences	Urban resilience refers to a system's capacity to withstand shocks, enhance its adaptive capabilities, and sustain normal functioning
Liao (2012)	Environmental sciences	Urban resilience refers to an urban system's capacity to minimize human casualties and economic losses, while sustaining normal functioning in the face of natural disasters like floods and earthquakes
Ernstson et al. (2010)	Social sciences/ environmental sciences	Urban resilience entails developing flexible and responsive capacities to address unpredictable risks and crises, ensuring the maintenance of a dynamic urban governance system
Campanella (2006)	Social sciences	Urban resilience refers to cities' capacity to recover from destruction
Almeida et al. (2020)	Engineering	Urban resilience entails the capacity to absorb, adapt to, and respond to changes in urban systems

(continued)

Table 1.4 (continued)

Authors	Research areas	Definition
Papa et al. (2015)	Environmental sciences	Urban resilience represents cities' endeavors to enhance their capacity to manage diverse stressors (e.g., climate, environment, energy, and economy) to promote a higher quality of life and sustainable urban development
Ribeiro and Gonçalves (2019)	Sociological	Urban resilience extends the concept of resilience from technical to social systems, encompassing the capacity to restore essential services and sustain citizens' livelihoods while preserving the core functions of the city amidst disasters
Desouza and Flanery (2013)	Sociological	Planning for urban resilience entails analyzing the impacts of urban pressures, assessing the city's vulnerability, understanding its processes and interactions, and building component structures to achieve resilience goals
Mendizabal et al. (2018)	Management studies	Urban resilience enables cities to absorb disturbances while maintaining essential functions, allowing the system to reorganize, adapt, and learn from disruptions
Balsas (2014)	Management studies	Urban resilience refers to the capacity of urban systems, including natural, human, and institutional aspects, to endure various changes, whether abrupt (shocks) or prolonged (crises), while preserving essential functions, structures, identities, and mechanisms, and evolving in a similar or revitalized fashion
Dongping et al. (2017)	Architecture	Urban resilience refers to the capacity of the entire urban system and its constituent subsystems to swiftly adapt and restore operational functions during crises, while also bolstering the system's adaptive capacity to ensure sustained development amidst potential future risks
Mingdou and Xiaoqing (2018)	Urban safety	Even when exposed to various disturbances and impacts from the external environment, the urban system can preserve its original structure, basic characteristics, and key operational components
Baoxing (2018)	Architecture	Urban resilience is a multifaceted concept categorized into three levels: structural resilience, process resilience, and system resilience. Structural resilience encompasses technical, technological, economic, social, and governmental resilience, while process resilience primarily pertains to the city's capacity to adjust, recover, and transform in response to shocks. System resilience addresses the overall resilience of the city as an open and complex adaptive system

(continued)

Table 1.4 (continued)

Authors	Research areas	Definition
Qihui (2017)	Urban safety	Urban systems demonstrate resilience, sustainability, and robustness in maintaining normal functionality amidst severe shocks and chronic risks
Jinhe and Hongxue (2020)	Urban economy	Urban resilience refers to the capacity of urban systems and their subsystems to identify, address, and proactively prepare for both existing and potential crises
Chong and Minxing (2017)	Science of disasters	Urban response and adaptation to both natural and man-made disasters are essential for ensuring urban resilience
United Nations "Making Cities Resilient" Program (https://www.preventionweb.net/publication/how-make-cities-more-resilient-handbook-local-government-leaders-2017)		A resilient city is capable of withstanding and absorbing the effects of disasters, shocks, and stresses through adaptation or retrofitting, thereby ensuring long-term sustainability and maintaining essential functions, features, and structures
William McCollum (https://iclei.org/e-library/resilient-cities-thriving-cities-the-evolution-of-urban-resilience-11/)		A resilient city is capable of maintaining its fundamental functions and structures while adapting and recovering from sudden crises or stresses. Moreover, it can identify opportunities for adaptation during crises and attain long-term sustainability. Strengthening resilience necessitates not only preemptively identifying potential risks but also mitigating the city's vulnerability and leveraging contingency and early warning measures
United States "A Stronger, More Resilient New York" program (Liping and Fengqi 2018)		The adaptive capacity of urban infrastructure systems, including transportation, energy, and communication systems, encompasses disaster impact reduction and the ability to swiftly restore city functions, thus fortifying cities against stress or shock events
Authority (2011)		Resilience refers to a city's capacity to maintain continuous resilience and be sufficiently prepared to handle future shocks

aspect of urban resilience is based on a vision for sustainable urban development, aiming to bolster cities' resilience to future uncertainties and achieve long-term urban development objectives through adaptation and transformation. This section synthesizes research and practice in resilient cities to define urban resilience as the ability of an urban system to rapidly adapt, recover, and transform in the face of various pressures and challenges, including natural disasters, climate change, and economic downturns. Urban resilience is primarily to establish flexible urban governance mechanisms, robust infrastructure systems, and the active participation and responsibility of communities and residents. These factors collectively contribute to sustainable urban growth and improve citizens' quality of life.

5. History of Urban Resilience

During the early stages of urban resilience research, scholars generally concentrated on urban resilience in the context of natural disasters, paying less attention to social aspects. The concept of "resilient cities" was first introduced in 2002 by the International Council for Local Environmental Initiatives (ICLEI). It aimed to strengthen research on safety resilience in urban disaster prevention (Motesharrei et al. 2016). The Hyogo Framework for Action (2005–2015) aimed to build the resilience of nations and communities to reduce disaster risks (Bing 2016). During this period, the primary emphasis was on governance for urban resilience, particularly in terms of amenities and environment management. In 2013, the Rockefeller Foundation started the Global Resilience 100 Cities Program to improve developmental resilience by addressing uncertain risks across natural, economic, and social dimensions (Spaans and Waterhout 2017). The sustainable development goals (SDGs) aim to enhance the safety, inclusivity, and resilience of cities and human communities (Guofang and Chenhong 2021). In 2016, the United Nations Conference on Housing and Sustainable Urbanization (UNCSU) proposed the New Urban Agenda, integrating "urban resilience" into the discourse on urban ecology and environment, aligning with the concept of sustainable urban development (Nan 2017). The United Nations Office for Disaster Risk Reduction (UNDRR) has actively addressed the DRR issues. Due to the consequences of the 2020 epidemic, the UNDRR's Making Cities Resilient 2030 program has prioritized resilience. The program will guarantee inclusive, safe, and sustainable cities by 2030. The program emphasizes the importance of resilience, striving for cities to achieve inclusion, security, resilience, and sustainability, signaling the increasing attention and support for the concept of urban resilience from government and academia.

Urban resilience has emerged as a prominent area of interest in urban safety, sparking enthusiasm for research in geography and related disciplines. International organizations like the Resilience Alliance, the Rockefeller Foundation, and UN-Habitat have also started prioritizing this topic. They have developed programs tailored to their specific resilience needs and implemented them in cities like London, United Kingdom, New York, and the United States. It has resulted in acquiring valuable knowledge and experience in urban resilience practices (Fig. 1.26).

6. Characteristics of Urban Resilience

Academia lacks a unified consensus on the characteristics of urban resilience. This section provides a summary of contemporary academic research on urban resilience characteristics and posits that urban resilience should encompass the following characteristics:

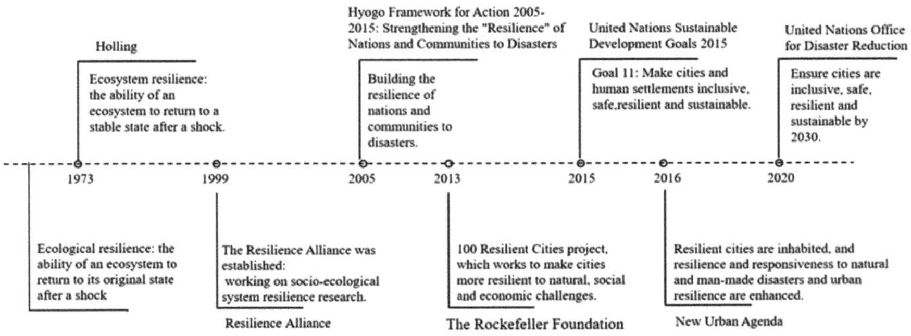

Fig. 1.26 Timeline of urban resilience research

(1) Robustness

Urban systems must have the ability to endure and absorb external disturbances of a specific magnitude while maintaining essential city functions, minimizing overall system losses post-perturbation (Xiaojun and Xin 2015). A city should not only effectively respond to emergencies like natural disasters, economic crises, and social unrest but also strive for sustainable development, enhanced quality of life, and relative stability amidst change. Urban robustness primarily guarantees cities' stability, adaptability, and rapid recovery amidst diverse internal and external pressures and challenges. Urban resilience relies heavily on the robustness of urban systems, which refers to their ability to endure disruptions and recover efficiently. Building urban robustness necessitates multifaceted efforts and strategies. Diversified infrastructure and resource distribution form a crucial foundation for urban robustness. Cities must establish diverse energy supply systems, transportation networks, water management systems, and communication systems to mitigate the impact of individual failures on urban functions. Flexible planning and management mechanisms are indispensable for enhancing urban robustness. Cities should possess robust emergency response mechanisms and crisis management systems equipped with flexible policies and regulations to address emergencies promptly. In addition, urban planning should consider future uncertainties and embrace resilient planning strategies to allocate sufficient space and resources for the city's development, enabling it to navigate potential future changes and challenges. Resilient social and economic systems constitute vital components of urban robustness. Cities must establish resilient job markets, adopt flexible business models, and strengthen social safety nets to address the impact of economic crises and social unrest on the lives of urban residents.

Building urban robustness is a continual process demanding cities to engage in perpetual learning and improvement. Cities can enhance their resilience and ability to adapt to future challenges and achieve sustainable development and urban governance goals by establishing diverse infrastructure and resource distribution, implementing flexible planning and management mechanisms, fostering resilient social and economic systems, promoting robust community participation and collaboration mechanisms, and improving learning and improvement capabilities.

(2) Efficiency

Urban systems should have effective scheduling capabilities, enabling cities to respond promptly to external disturbances by swiftly identifying their origins and assessing their impact (Yaru 2023). The high efficiency of urban systems is a crucial aspect of urban resilience, primarily manifested in optimizing resource allocation, rapid response capacity, efficient public service provision, and flexible decision-making mechanisms. These characteristics can enhance cities' competitiveness, enhance citizens' living quality, and enable effective adaptation to challenges and changes. High efficiency holds significant importance for urban resilience, as the efficient operation of urban systems can furnish cities with additional resources and support during responses to external perturbations and shocks. It ultimately strengthens the resilience capacity of urban systems. Moreover, the efficient operation of urban systems can mitigate resource loss and waste, thereby fostering the sustainable development of cities.

(3) Redundancy

Redundancy in urban systems is a critical aspect of urban resilience. Urban systems should assign extra capacity beyond their immediate requirements to reduce sudden shocks and distribute potential risks. It will enable them to quickly adjust to external perturbations and changes (Yiwen and Jiang 2015). Redundancy significantly enhances the robustness and adaptability of urban systems, mitigating the risk of system-wide failure stemming from single points of failure. Thoroughly considering and implementing redundancy in urban planning and construction is crucial. Urban systems should implement measures to enhance infrastructure redundancy and spatial layout redundancy. For instance, multiple redundant urban arterial roads and public facilities can be constructed to ensure the continued provision of essential social services in case of the failure of a single facility. Redundancy in the city's spatial layout is crucial for its sustainability, and allocating space for future growth enhances the city's resilience. Thus, constructing multiple alternative transportation nodes and dispersing traffic pressure in planning can enhance the resilience of the urban transportation system. Moreover, redundancy is vital to enhancing urban resilience and should be thoroughly considered and integrated into urban planning and construction. Enhancing the urban system's redundancy ensures the urban system's stable operation and sustainable development, improves the city's adaptability to external disturbances, and realizes the goal of sustainable development.

(4) Diversity

At the city-wide level, the urban system effectively manages external disruptions by employing various coping strategies devised by different service sectors inside the city. At the individual level, communities, enterprises, governmental bodies, and other diverse stakeholders can exchange information and facilitate resource mobilization to collectively address disruptive shocks (Ahern 2011). Urban resilience manifests diversity through various facets, including infrastructure, communities, economic structures, and cultural and talent diversity. Urban infrastructure diversity is a vital component of urban resilience. Infrastructure systems of varying types and scales comprise the backbone of urban operations, encompassing transportation, energy, water, and communication networks. Diversity within urban communities is also integral to urban resilience. Cities have multiple communities with distinct traits, cultural backgrounds, and social organizational forms. Furthermore, the diversity of the city's economic structure plays an essential role in supporting urban resilience. The city's economy covers various industries and sectors, such as manufacturing, services, and finance. The diversity of cultures and talents within cities is another critical aspect of urban resilience. Urban resilience diversity serves as a crucial safeguard and support for urban development. It bolsters the city's resistance and adaptability to external threats, mitigates the impact of singular factors on urban development, increases the resilience level of the urban system, and fosters the city's sustainable development.

(5) Adaptability

Adaptability refers to the capacity of the urban system to learn from various types of shocks, incorporating lessons and innovating through active or passive learning. It enhances the city's ability to resist, adapt, and recover, thus better managing potential risks in the future (Allan and Bryant 2011). It reflects the resilience and adaptability of cities in a dynamic environment, constituting a key factor for cities to address both external disturbances and internal changes. The adaptability of urban resilience is evident in the flexibility of urban planning and construction processes. Urban planning and construction must consider future uncertainties and changes, adopting flexible designs and strategies to accommodate the dynamic changes in urban development. Moreover, the adaptability of urban resilience is demonstrated through the flexibility of urban infrastructure and public services. Urban infrastructure is a crucial support for city operations, while urban public services are essential for ensuring residents' well-being and safety. In response to unforeseen events such as natural disasters, technological failures, and human-induced damage, cities must swiftly adjust to take emergency measures and operate flexibly to ensure the stable operation of infrastructure and the uninterrupted provision of public services. For instance, following a natural disaster, cities can promptly restore damaged infrastructure and services using temporary facilities and emergency stockpiles to ensure the safety and well-being of residents. The adaptability of urban resilience extends to the flexibility of the urban economic structure and industrial layout. Urban social organization and governance mechanisms serve as vital safeguards for urban operation and management, with urban residents constituting an integral part of urban society. Under the challenges of social change, population mobility, and cultural conflict, cities must be capable of swift adjustment and adaptation to ensure stability in social organization and the effective functioning of social order. Therefore, in urban management and development, efforts should be directed towards enhancing the adaptability of urban resilience, bolstering urban planning and design, fostering technological innovation and organizational management, and enhancing the overall adaptive and coping capacity of cities (Fig. 1.27).

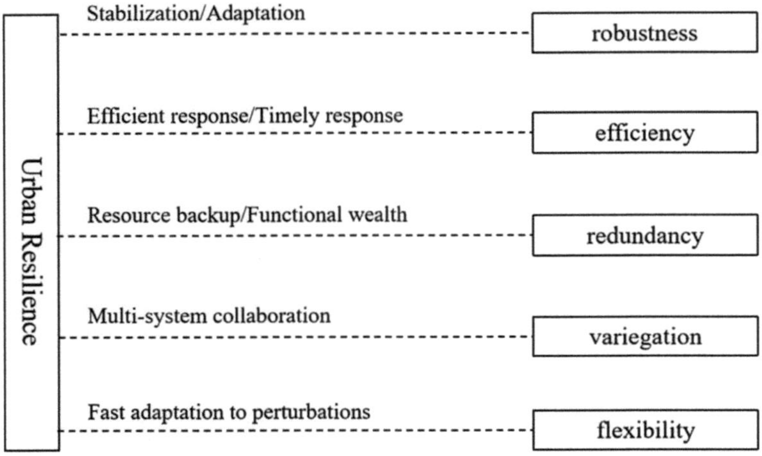

Fig. 1.27 Urban resilience characteristics

Urban Safety Resilience

Since its emergence, the concepts of resilience and urban resilience have proven to be very applicable across various research disciplines, aims, and contexts. The academic community's comprehension of urban safety resilience has progressively evolved.

Bibliometric Analysis Related to Urban Safety Resilience

An analysis of 49 CSSCI articles on "urban safety resilience" in the CNKI database from 2015 to 2023 with a one-year time slice reveals notable contributions from terms such as urban safety, urban resilience, resilience, safety development, disaster prevention, and mitigation, as shown in Fig. 1.28.

Analysis of the time zone view of urban safety resilience reveals that research commenced relatively late in the CNKI database, with its origins traceable back to 2015 in the existing literature. Initially, the database focused its research on the resilience of cities as a major subject. As time progressed, the research direction of urban resilience gradually expanded to encompass the realms of urban safety, disaster prevention and planning, disaster prevention and mitigation, urban governance, and community autonomy. It is evident that urban security resilience, smart cities, and urban security gradually emerge as prominent trends and directions of research, as shown in Fig. 1.29.

Citation bursts of urban safety resilience reveal the presence of "disaster prevention and mitigation," "resilient community," "resilience," "public safety," and

Fig. 1.28 Co-occurrence network of urban safety resilience keyword in the CNKI database

Fig. 1.29 Time zone view of urban safety resilience keywords in the CNKI database

Fig. 1.30 Citation bursts of urban safety resilience in the CNKI database

"resilient governance." Although research on themes connected to these keywords has emerged relatively late, they nonetheless represent prominent research hotspots and trends. Evidently, research on urban safety and resilience has evolved into a theme focused on coupling and coordination within urban governance, as shown in Fig. 1.30.

In summary, the results of the bibliometric study indicate a shift in research focus from initial investigations into urban resilience and urban safety towards resilience within governance and related domains. The research on urban safety resilience has remained consistent and comprehensive. Furthermore, the research direction of urban safety resilience in the CNKI database has progressively shifted towards investigating coupling and coordination with other facets. Coupled and coordinated research on urban safety resilience is expected to become increasingly prominent.

Literature Review on Urban Safety Resilience

1. Research on the Problems of Urban Safety Resilience

Recognizing the advancements achieved in urban safety resilience is essential, but it is also necessary to acknowledge the numerous persistent challenges and issues.

(1) Issues Affecting the Development of Economic Security Resilience

Economic security resilience is essential for sustainable development and stability of cities. Economic security resilience refers to the ability of the metropolitan economic system to maintain stability, adaptability, and resilience, which can reduce economic losses and ensure social stability. Improved economic security resilience enhances a city's competitiveness and resilience to risks, attracts investment, promotes employment, improves people's livelihoods, and lays a solid foundation for long-term prosperity and development.

Therefore, special attention needs to be paid during the development process. Insufficient economic vitality in a city and the level of urban economic security directly impact its economic activity. The security level is a significant factor that limits the city's economic development. Urban economic vitality is the capacity to sustain the development of the urban economy and encompasses five main characteristics: resilience, agglomeration, innovation, market dynamism, and radiation (Xiaoguang and Ying 2022). From a macro-perspective, scientific and technological innovation, participation of social organizations, market mechanisms, and government support are the primary factors influencing economic vitality (Xiaoli and Wanshan 2007). From a micro-perspective, financial development, efficiency in capital allocation, and Internet penetration rate also play crucial roles in economic development. Insufficient vitality in a city's economic development will significantly impact its healthy and sustainable growth (Zhen et al. 2021). For instance, a lack of economic vitality may lead to the city losing its attractiveness and failing to attract sufficient investment and talent inflows, which limit its economic growth potential. This situation may also result in a decline in the competitiveness of urban enterprises, a concentration on a single industrial structure, and difficulty adapting to external shocks and market changes, which will increase the risks and uncertainties. Additionally, the lack of economic vitality will impact the quality of life for residents, leading to challenges such as limited employment opportunities, decreased earnings, and inadequate social welfare support.

Besides, the frequency of emergencies affecting cities also impacts economic security resilience. There are some extreme accidents that threaten economic security: the "Xiangshui Chemical Plant Explosion" in 2021 (https://www.gov.cn/xinwen/2019-11/15/content_5452468.htm), the "Collapse of Xinjia Express Hotel" in 2020 (https://news.sina.com.cn/o/2020-12-29/doc-iiznezxs9580721.shtml), and the "Beirut Explosion" in 2020 (https://www.thepaper.cn/newsDetail_forward_10923560). These incidents result in significant human and property losses and instigate anxiety and social unrest among residents, detrimental to urban social stability. These incidents reflect cities' imbalanced pursuit of economic benefits, disregarding social and security concerns. These incidents have resulted in the temporary stagnation of economic activities and decreased productive capacity, leading to economic losses and heightened market uncertainty. Simultaneously, they have prompted cities to bolster the development of economic security resilience. It includes enhancing contingency plans, strengthening supply chain management, and improving the resilience of industries. These efforts aim to enhance cities' ability to manage external risks and shocks, fostering the development of security resilience in urban areas.

(2) Issues Affecting the Development of Urban Ecological Security and Resilience

The significance of ecological security resilience cannot be underestimated. Ecosystems serve as the foundation for sustaining human social and economic development. They encounter a multitude of challenges, including climate change, environmental pollution, and biodiversity loss. Effective ecological security resilience entails ecosystems maintaining their function and stability despite external shocks. This protects the ecological environment, preserves biodiversity, and maintains ecological balance.

Urban development has brought about severe pollution and ecological deterioration. The city's air quality is greatly affected by factors such as automobile exhaust, industrial emissions, and coal combustion. It has resulted in severe atmospheric pollution, exceeding standards for PM2.5 and other harmful substances. Between 2009 and 2016, the concentration of PM2.5 in the Sichuan-Chongqing urban agglomeration ranged from 61.3 to 165.1 μg/m^3, with an average of 102.5 μg/m^3. In 2012 and 2013, the winter concentrations of PM2.5 in Chongqing and Chengdu were higher than 90.1 μg/m^3 and 108.1 μg/m^3, respectively. Chongqing and Chengdu had five and nine days of heavy pollution, with PM concentrations exceeding 150 μg/m^3 (Shengdong et al. 2020). Besides, environmental issues impact the resilience of urban ecological security. Additionally, there is the issue of disorderly expansion in urban land use: certain local governments, aiming for GDP growth, have converted significant amounts of arable land into construction sites, accelerating urbanization without proper planning (http://theory.people.com.cn/n1/2016/0801/c217905-28601100.html).

Adequate land development is a crucial component of sustainable urban development. In urban governance, the interplay between economic development and environmental protection must be comprehensively addressed to achieve economic and ecological benefits. Enhancing the total security resilience of cities can only be achieved by taking this action.

(3) Issues Affecting the Development of Urban Cultural Security and Resilience

When striving to improve the resilience of urban cultural security, there are difficulties in elevating the level of cultural security in cities. First, inadequate protection of cultural heritage is at risk due to various issues like globalization, urbanization, increased tourism, the smuggling of cultural relics, atmospheric pollution, population growth, wars, and natural disasters. The Taliban destroyed the Bamiyan Buddhas in Afghanistan in the name of "opposing idolatry" (https://uci.sdu.edu.cn/info/1219/1311.htm). Second, there is a loss of cultural distinctiveness in cities, with many becoming homogeneous and losing their unique cultural identity. Third, there are issues with establishing a safety culture, which lacks synergy, theoretical guidance, standardization, and long-term mechanisms and requires transformation (file:///C:/Users/Ling/Zotero/storage/ZD4JN7P7/t20210202_2251256.html).

Strengthening urban safety resilience necessitates close cooperation among all sectors of society, and a healthy and positive culture of urban resilience must permeate throughout society. Cultivating a culture of urban cultural safety resilience can enhance urban residents' risk awareness and resilience, enabling them to effectively respond to disasters and recover promptly, thereby bolstering the cohesion and sense of security within urban communities.

2. Research on Factors Influencing Urban Safety Resilience

The study of urban safety resilience examines cities' ability to manage disasters and challenges and thoroughly explores the factors influencing urban safety resilience. Factors influencing urban safety resilience encompass the economic foundation, social conditions, administrative capability, and technological advancement of cities. Moreover, factors such as geographic location, level of economic development, diversity in industrial structure, the willingness of cities to collaborate, market openness, disparities in political resources, and the extent of innovation exchange also play a significant role in shaping urban security resilience. Most scholars assess the condition of urban systems by examining the composition of urban system elements, primarily analyzing the impact of material elements (i.e., infrastructure, ecology) and non-material elements (i.e., economy, society, organization) on urban resilience (Shi et al. 2021).

(1) In Terms of Material Elements:

Infrastructure is crucial in supporting urban disaster resilience, constituting the frontline defense against disasters. It is crucial in ensuring the urban system's resilience and vitality and facilitating residents' normal production and life. Urban infrastructure primarily consists of urban buildings, transportation facilities, public infrastructure, and communication facilities (Longyu et al. 2022). Zhang Sisi argued that population agglomeration positively enhances urban resilience, with human capital partially mediating this relationship (Sisi et al. 2023). The road transportation network is a vital component of the urban critical infrastructure system (Yuanbing et al. 2023), playing a crucial role in disaster rescue efforts, with better road transportation performance facilitating more effective rescue operations (Zhang et al. 2021b). The city's safety resilience is enhanced. The impact of the ecological environment on urban safety resilience cannot be overstated, given its importance in regulating natural disasters, providing resources and ecosystem services, enhancing residents' health and quality of life, and maintaining ecological balance and biodiversity. The ecological environment serves as the spatial foundation for urban sustainable development, encompassing the regional natural environment and urban landscape green space. It can provide diverse ecosystem services based on natural ecological processes to enhance urban ecological resilience (Ruidong et al. 2020). Ecological environment construction is widely acknowledged for its role in reducing flood risk, alleviating the urban heat island effect, enhancing air quality, and preserving biodiversity. It has become a crucial component of urban resilience strategy in urban planning (Meerow and Newell 2017).

(2) In Terms of Non-material Elements:

Considering the economic aspects, it is widely acknowledged that cities with higher levels of economic development tend to possess diversified industrial structures that foster technology innovation. Residents in these cities also have more benefits and opportunities for employment, education, services, and skills training (Jieyao et al. 2020). Having a diverse industrial base is crucial for economic recovery, and cities can effectively respond to changes in industrial structure by investing in diversification and balancing support for existing industries with the promotion of new ones. Besides, adopting new concepts and advanced technologies such as big data, the Internet of Things, cloud computing, and artificial intelligence enables the acquisition, simulation, analysis, and processing of information before disasters occur. It improves government and individual capabilities for post-disaster emergency communication, security, risk and loss assessment, emergency response, disposal, restoration, and reconstruction, thus enhancing the city's disaster-coping abilities. Digital technology development fosters interpenetration and integration among the city's various subsystems. The optimization of public service allocation and resource utilization, such as urban roads, education, and health care, not only enhances service redundancy but also facilitates the transition from old to new energy sources. It improves urban operational efficiency and comprehensively boosts the urban system's overall performance, enhancing security resilience against unknown risks and further fortifying the city's ability to withstand disasters (Bin et al. 2023). Thus, urban security resilience is influenced by factors such as the city's economic status and economic development.

Considering the social environment aspects, urban areas epitomize humans as social beings. Cities are ideal environments for fostering social connections, building networks, and cultivating a sense of community. Urban areas are distinctive due to the proximity of individuals with diverse norms, cultures, and backgrounds, resulting in multi-layered communities and complex networks (Kapucu et al. 2021). Ribeiro identifies urban neighborhoods, community innovation initiatives, and habits as key social factors influencing urban resilience (Ribeiro and Gonçalves 2019). Meerow argued that social equity significantly influences urban resilience (Meerow et al. 2019). The social environment encompasses crucial areas such as health care, education, economic capital, social justice, population dynamics, employment opportunities, and social security. These components collectively contribute to the resilience and adaptability of cities.

Considering the organizational management aspect, it plays a crucial role in urban safety resilience. Fastiggi et al. investigated the impact of urban resilience governance by analyzing the organizational structure of 20 governments in North America (Fastiggi et al. 2021). Huck et al. noted that a single managerial, organizational structure impedes the efficient collaboration of a multi-level governance network for urban resilience. They emphasized that collaboration among multi-tiered sectoral and organizational networks is crucial for modern urban emergency governance and for implementing urban resilience. Horizontal organizational networks enhanced the effectiveness of cross-sectoral problem-solving (Longyu et al. 2022). Effective crisis management, resource allocation, cooperation and coordination, information sharing, and community participation in organizational management for urban safety resilience can enhance the city's ability to respond to diverse challenges and safeguard the lives and properties of urban residents.

3. Research on Urban Safety Resilience Response Strategies

Urban safety resilience is widely recognized to impact a city's development positively. Given the rising frequency of natural disasters and man-made risks, urban safety resilience has become an urgent priority for contemporary urban management. Effective coping strategies depend on scientific early warning systems, emergency response mechanisms, and comprehensive measures such as community participation, multidimensional governance, infrastructure resilience, intelligent urban design, and public education. These measures aim to build safer, more robust, and adaptable urban environments.

(1) Increasing the Resilience of Urban Economic Security

Promote economic diversification to reduce reliance on specific sectors or critical industries, mitigating the impact of specific risks on the city's economy. Invest in and enhance urban infrastructure, particularly in transportation, energy, and communications sectors, to bolster the city's resilience against external disruptions. Promote digital transformation and support the growth of cutting-edge industries and novel economic models to enhance the adaptability and resilience of urban economies. Establish a flexible labor market, reinforce vocational skills training and job transition services, and enhance the adaptability of urban residents to economic changes. Strengthen financial regulation and risk management to mitigate the impact of financial risks on the urban economy, ensuring the soundness and sustainability of the financial system. Enhance urban governance capacity and public service provision to bolster the city's ability to respond to economic risks and manage crises effectively. Enhance international economic cooperation and exchanges to collectively tackle global challenges, share experiences and resources, and bolster the competitiveness and resilience of cities in the international market.

(2) Increasing the Resilience of Urban Ecological Security

Develop and implement ecological strategies to protect and restore natural ecosystems within cities, such as wetlands, forests, and grasslands, to enhance urban areas' ecological diversity and stability. Promote sustainable urbanization by enhancing energy efficiency, reducing carbon emissions, and minimizing pollutant emissions to alleviate pressure on the ecological environment. Enhance urban greening and ecological landscape design to expand green spaces and corridors, fostering improved ecological connectivity and biodiversity within urban areas. Improve the management and protection of water resources, encompassing water source protection, purification, rainwater collection, and utilization, to enhance the sustainable use of water resources in cities. Advance the development of a circular economy to minimize resource consumption and waste emissions, enhance resource utilization efficiency, and lessen the environmental burden on cities. Conduct ecological and environmental education to raise public awareness of and engagement in ecological and environmental protection, fostering residents' awareness and capacity to engage in environmental conservation. Strengthen monitoring, early warning, and response to ecological disaster risks, and implement effective measures to mitigate the losses and impacts of ecological disasters on cities.

(3) Enhancing the Resilience of Urban Cultural Security

Strengthen the protection and conservation of urban cultural heritage, including historical buildings, traditional crafts, and folk culture. Policy and programmatic support ensures their continuous existence and growth in urban development. Foster the development of cultural and creative industries by supporting cultural enterprises and creative talents, nurturing cultural and creative industry clusters, and boosting the city's cultural innovation capacity and economic vitality. Design and strategize urban public layouts to cultivate urban environments with distinct cultural attributes and liveliness, promoting cultural exchanges and interactions to enhance the sense of belonging and cohesion among urban residents. Advocate for multicultural exchanges and integration, supporting collaboration among diverse cultural groups to promote cultural diversity and inclusivity, fostering a harmonious and coexisting urban society. Strengthen cultural education and outreach efforts to enhance citizens' cultural literacy and self-confidence, fostering urban residents' awareness of and engagement in culture. Enhance disaster protection and response capacity for cultural heritage, implementing measures to safeguard cultural heritage from natural disasters and human-induced damage, ensuring the sustainable development of urban culture. Formulate and implement urban cultural policies and plans, elucidating the guiding ideology, goals, and pathways of cultural development to ensure cultural pursuits' sustainable and healthy growth.

4. Urban Safety Resilience Evaluation Study

The prevailing trend in assessing urban security resilience still revolves around urban resilience research, which integrates security considerations to evaluate urban security resilience.

(1) Urban Safety Resilience for Different Types of Disasters

The predominant focus of research on urban safety resilience in the CNKI and WOS database pertains to natural disasters. Chen Peng examined the impacts of urban flooding disasters on residents, society, and the economy in Harbin City, using a typical year of flooding disaster as a case study to investigate urban safety resilience in flood scenarios (Peng et al. 2015). Sun Hongguang explored the human-earth relationship implications of urban haze disaster resilience from the perspective of residents' activities, conducting an empirical analysis of resilience assessment in the main urban area of Nanjing City and proposing targeted resilience enhancement strategies (Honghu and Feng 2019). Yu Hongxia introduced the concept of "initiality" regarding urban and seismic toughness levels of buildings, elucidating the limitations of traditional design methods and the advantages of toughness-based seismic design methods through comparative analysis (Xue et al. 2018).

(2) Urban Safety Resilience for Different Types of Cities

In addition to evaluating safety resilience based on types of disasters, there is also evaluation based on different types of cities. Li Min developed an evaluation model for safety resilience cities by studying a mega city's response to natural disasters (Min 2021). Zhang Beibei established an evaluation system with 24 indicators across five dimensions: economy, culture, infrastructure, society, and ecology. They assess the resilience levels of 41 cities in the Yangtze River Delta to analyze the temporal and spatial characteristics (Beibei et al. 2023). Zhao Li utilized the urban linear model to construct an urban resilience evaluation index system for the Beijing-Tianjin-Hebei region. These scholars analyze various city types, contributing to a comprehensive understanding of urban safety resilience (Li et al. 2023).

The assessment of urban safety resilience extends beyond applied research on various disasters or city types, with notable advancements in theoretical research approaches.

(3) Qualitative Research on Urban Safety Resilience

Wu Bohong analyzed the evolutionary mechanism of resilient cities encountering external perturbations. They proposed a resilience evaluation model to identify critical nodes' social significance in resilience evolution (Bohong and An 2018). Xu Zhaofeng examined urban disaster resilience across infrastructure, economic, social, and organizational/institutional aspects, constructing an evaluation index system. They developed a comprehensive evaluation model using hierarchical analysis to weigh each index (Zhaofeng et al. 2019). Huang Hong et al. explored the concept of safety resilience cities, analyzing its origin and connotation, and proposed a model for characterizing urban safety resilience, emphasizing essential city attributes (Hong et al. 2018b). These scholars have considered internal and external city diversity, community engagement, governance mechanisms, infrastructure quality, and residents' coping capacity, aiming to comprehensively assess cities' performance in disaster and challenge response.

(4) Quantitative Research on Urban Safety Resilience

Jiao Liudan developed an urban resilience evaluation index system with 43 indicators based on the pressure-state-response framework, further creating an evaluation model using the entropy weight method and cloud model (Liudan et al. 2022). Zhang Heng utilized a GA-BP neural network model to analyze the spatiotemporal evolution trend of city safety toughness, offering recommendations to enhance overall safety resilience (Heng et al. 2024). Liu et al. employed simulation methods to probabilistically assess safety toughness levels under urban flooding disasters (Liu et al. 2020). Quantitative analysis involves gathering extensive data and assessing indicators such as urban losses, recovery speed, and the efficacy of countermeasures. This process allows for measuring cities' performance in dealing with disasters and challenges, providing an objective basis for policy formulation and enhancing urban safety resilience.

Review of Research

There has been an increase in the number of keyword research findings about urban safety resilience. Scholars have mainly concentrated on urban safety, disaster prevention and planning, disaster prevention and mitigation, urban governance, and community autonomy. There are relatively few studies directly addressing urban safety resilience. Due to the growing complexity of environmental and societal threats that cities encounter, examining urban safety resilience has become paramount. Despite scholars' efforts in the CNKI and WOS databases, there is still a lack of comprehensive and practical study in this field. Given the increasing urban disaster risks, enhancing urban safety resilience has become imperative for ensuring urban safety and sustainable development. The core lies in effectively responding to various changes, shocks, or disturbances to reduce instability and vulnerability in urban development. The research on urban safety resilience still exhibits several blind spots and deficiencies, as outlined below:

(1) Problematic Aspects: Although the concept of urban safety resilience has been introduced, its definition remains ambiguous and diverse, resulting in biases and disparities in research understanding. Scholars in the database predominantly explore economic, social, ecological, and infrastructure resilience separately, lacking a comprehensive understanding of urban safety resilience.
(2) Influencing Factors: Although several factors influencing urban safety resilience have been identified, such as urban planning and design, socio-economic factors, and government governance and management, a comprehensive system of influencing factors has not yet been established. Insufficient comprehensive study exists on the interactions between these factors and an analysis of their weighting on urban safety resilience.
(3) Coping Strategies: Existing coping strategies primarily emphasize economic security resilience, ecological security resilience, and cultural security resilience, but their practical implementation faces limitations. There is a dearth of thorough assessments regarding the effectiveness and feasibility of strategy implementation and the development and implementation of tailored strategies for different city characteristics.
(4) Evaluation: Despite its importance in urban safety resilience research, evaluation lacks a unified, comprehensive, and scientifically sound set of methods and indicator systems. Diverse research methods hinder measurement and comparison, affecting the credibility and comparability of outcomes, thereby limiting practical guidance for enhancing urban safety resilience.
(5) Interdisciplinary Research: Urban safety resilience involves various disciplines, yet research often lacks integration due to scholars approaching the topic from their disciplinary backgrounds. This fragmentation hinders a comprehensive understanding and analysis of urban safety resilience.

The various shortcomings have led to a broader distribution of research findings and viewpoints on urban safety resilience. The absence of a comprehensive and systematic research framework has made it difficult for subsequent researchers to comprehensively understand and respond to cities' complex challenges and disaster risks. To better address these issues in subsequent research, interdisciplinary cooperation and communication must be strengthened to enhance urban safety resilience. Hence, the coupled and coordinated development of urban safety resilience is an important topic that needs to be addressed.

The Concept of Urban Safety Resilience

According to the UNISDR, urban safety resilience refers to the ability of a system, community, and society exposed to hazards to withstand and adapt to change, sustaining an acceptable level of functionality (Sendai Framework for Disaster Risk Reduction 2015–2030). Merrow defined urban safety resilience as the capacity of urban systems to preserve desired functions or swiftly recover, coupled with the capability to adapt to change through rapid transformation (Meerow et al. 2016). Zhou Limin characterized a safe and resilient city by its ability to rapidly reorganize and restore daily life and production following disaster shocks (Limin 2016).

Urban safety resilience has evolved from a singular focus on engineering resilience to a more integrated, ecological, and evolutionary perspective. Urban safety resilience now emphasizes emergency response and post-disaster repair and the overall health, stability, dynamic interaction, and adaptation of the urban system with its environment. This evolution reflects a deeper and broader understanding of urban safety resilience, offering a more expansive vision and theoretical foundation for future research and implementation.

Characterization of Urban Safety Resilience

With the rapid development of cities, they face a growing number of security challenges. As complex, interconnected systems, cities rely on their security resilience to ensure the continuity of essential functions in the face of disruptive events. Urban safety resilience shares many characteristics with resilient systems and typically exhibits the following attributes:

First, resilience pertains to the capacity of a system and its components to uphold normal system function in the face of shocks. Urban safety resilience underscores the urban system's capability to preserve relative integrity amid disaster, allowing the city to sustain normal operations and safeguard critical functions from damage.

Second, redundancy denotes the extent of substitutability among the components of a system. Redundancy enables the system to maintain relative stability of its functions when specific components fail due to external shocks by utilizing additional resources. In the urban context, redundancy entails having backup facilities for essential functions of the urban system. These backup facilities can assume the corresponding functions when primary urban facilities suffer destructive damage. The measure prevents interruptions in urban system functions, ensures smooth operation, and reduces damage caused by disturbances.

Third, efficiency denotes the system's capacity to promptly achieve its objectives when confronted with a security risk, reducing the damage resulting from a disaster. Efficiency in the urban system requires a prompt response from the city's chain of command, which is imperative when faced with a security crisis. It involves promptly establishing early warning systems, organizing command and control scheduling, and executing measures to mitigate consequences during the initial stages of the incident.

Fourth, diversity denotes the varied responses of subsystems within the metropolitan system to identical shocks. Given the complexity of the urban system, it comprises numerous subsystems such as the economic, ecosystem, and political systems. Each subsystem responds uniquely to external shocks, contributing to the intricate process through which the urban system restores equilibrium.

Fifth, self-organization highlights a system's capability to manage risk through internal mechanisms. Self-organization is facilitated by individuals who orchestrate interventions aimed at swiftly restoring city functions. These interventions encompass organizational management, communication, collaboration, and other strategies.

Sixth, learning pertains to a system's capacity to assess its effectiveness in managing external shocks, encompassing the ability to prioritize and acquire diverse solutions. Learning entails the city's capability to efficiently utilize diverse resources and leverage its learning function to respond to shocks. It involves making enhancements for future shocks and enhancing the urban system's adaptability to uncertain shocks.

Seventh, wisdom denotes a system's capacity to optimize its response to shocks and adapt to external challenges through wise optimization. In the context of urban systems, intelligence is manifested in optimizing urban security emergency management systems. It includes accumulating response experience, analyzing it through collection, aggregation, feedback, and storage, and optimizing the urban system to enhance its response to future security incidents.

Components of Urban Safety Resilience

Cities are fundamental units of human habitation, comprising diverse elements. Bruneau scholars defined urban components as encompassing a broad spectrum of tangible and intangible elements, ranging from natural and social factors to institutional design, public behavior, and organizational effectiveness, including engineering capacity (Weicheng 2020). The elements within a city can be categorized into four dimensions: engineering, social, economic, and organizational aspects. By integrating system theory with cities' characteristics, security resilience can be defined as the ability of the city's subsystems across different dimensions to maintain stability, withstand shocks, and adapt to various crises in the face of security events. The following section delineates the four dimensions of security resilience.

(1) The engineering dimension pertains to the physical security infrastructure essential for the city's survival and development, encompassing roadways, communication networks, water supply, drainage systems, and other related components. The dimension is also known as the infrastructure dimension. Engineering resilience constitutes one of the components in urban safety resilience. When

external disturbances arise, the inherent anti-disturbance capacity of urban infrastructure and interdependencies among urban infrastructure elements can mitigate damage, curb the chain effects of urban disasters, and decrease the likelihood of secondary disasters.

(2) The social dimension comprises two primary levels: governmental institutions and public/non-governmental organizations and institutions. Resilience at the social dimension level primarily reflects the impact of societal actors on urban resilience, with social resilience primarily offering intangible support for urban safety resilience. Positive feedback from governmental, public/non-governmental organizations and institutions during and after a disaster can mitigate its adverse effects.

(3) The economic dimension significantly influences urban safety and resilience construction. The status of urban economic development and economic capacity directly impacts the scale and sustainability of future urban growth. After a disaster, economic support is required for rescue efforts and post-disaster reconstruction. Moreover, a resilient economic structure is essential for minimizing economic losses resulting from the disaster. Effective urban economic resilience can regulate, mitigate, and alleviate the direct impacts of disaster disturbances. Additionally, it can reduce disaster losses by concurrently providing relief materials, post-disaster economic subsidies, and compensation.

(4) The organizational dimension constitutes another intangible aspect of urban safety resilience. The city's organizational resilience manifests in its capacity to effectively utilize various urban elements to adapt to disasters, bolstered by its organizational prowess. The resilience also augments the city's ability to preempt and mitigate disasters through adaptive measures. Improving urban organizational resilience involves not altering the objective conditions of the city but instead organizing a series of disaster prevention and reduction education and training initiatives, bolstering the effective utilization of urban informational resources, studying urban emergency response experiences, optimizing the city's organizational structure, and employing other strategies to enhance the city's organizational resilience and reduce disaster losses.

Cities should demonstrate resilience and successfully reduce the loss of human life and property caused by disasters, considering the four dimensions of engineering, society, and economy. The different components within the city are interconnected and interact, collectively forming the functional essence of a safe and resilient urban environment. According to resilience theory, a safe and resilient city should possess disaster resilience, tolerance, self-recovery capability, and disaster learning ability. It is crucial to maintain relative stability inside the urban system even when it is affected by external factors. This will ensure that critical functions continue to operate while minimizing disaster losses and impacts. After a disaster, the city should possess the adaptive capacity to resume normal production and life swiftly, assimilate lessons learned, and make adaptive adjustments to enhance disaster prevention and reduction capabilities. The fundamental departure from traditional disaster prevention and reduction approaches is that a safe and resilient city prioritizes not merely responding to disasters but leveraging and managing disasters while prioritizing safety. The primary target of urban construction shifts towards utilizing and safeguarding against disasters, thereby ensuring the city's capability to address disasters of varying magnitudes (Enxin 2016).

References

2017. *How to Make Cities More Resilient: A Handbook for Local Government Leaders.* https://www.preventionweb.net/publication/how-make-cities-more-resilient-handbook-local-government-leaders-2017

Adger, W. N., T.P. Hughes, C. Folke, et al. 2005. Social-ecological resilience to coastal disasters. *Science* 309 (5737): 1036–1039.

Ahern, J. 2011. From fail-safe to safe-to-fail: Sustainability and resilience in the new urban world. *Landscape and Urban Planning* 100 (4): 341–343.

Aitsi-Selmi, A., V. Murray, C. Wannous, et al. 2016. Reflections on a science and technology agenda for 21st century disaster risk reduction: Based on the scientific content of the 2016 UNISDR science and technology conference on the implementation of the Sendai framework for disaster risk reduction 2015–2030. *International Journal of Disaster Risk Science* 7 (1): 1–29.

Alberti, M., J. M. Marzluff, E. Shulenberger, et al. 2003. Integrating humans into ecology: Opportunities and challenges for studying urban ecosystems. *BioScience* 53 (12): 1169–1179.

Allan, P., and M. Bryant. 2011. Resilience as a framework for urbanism and recovery. *Journal of Landscape Architecture* 6 (2): 34–45.

Almeida, M. do C., M. J. Telhado, M. Morais, et al. 2020. Urban resilience to flooding: Triangulation of methods for hazard identification in urban areas. *Sustainability* 12 (6): 2227.

Authority, G. L. 2011. *Managing Risks and Increasing Resilience. The Mayor's Climate Change Adaptation Strategy.* City Hall, The Queen's Walk, London, UK.

Ayers, J., S. Anderson, S. Pradhan, et al. 2012. *Participatory Monitoring, Evaluation, Reflection and Learning for Community-Based Adaptation: A Manual for Local Practitioners.* Atlanta, USA: CARE International.

Balsas, C. J. 2014. Downtown resilience: A review of recent (re)developments in Tempe, Arizona. *Cities* 36: 158–169.

Baoxing, Q. 2018. Methods and principles of designing resilient city based on complex adaptive system theory. *Urban Development Studies* 25 (10): 1–3 (in Chinese).

Beibei, Z., Z. Dongyue, and L. Yan. 2023. Research on temporal and spatial evolution characteristics of urban resilience in Yangtze River delta region. *Journal of Liaoning University of Technology (Natural Science Edition)* 43 (5): 325–331 (in Chinese).

Berkes, F., and C. Folke. 1994. *Linking Social and Ecological Systems for Resilience and Sustainability.* Beijer International Institute of Ecological Economics.

Bin, S., and H. Lu. 2021. Review and prospect of urban disaster prevention and reduction planning. *Huazhong Architecture* 39 (11): 86–91 (in Chinese).

Bin, Z., C. Binsen, W. Xiaojing, and W. Chuanhao. 2023. Research on the influence mechanism of digital economy on urban resilience: Taking the Yangtze River delta urban agglomeration as an example. *Journal of Chongqing University of Arts and Sciences (Social Sciences Edition)* 42 (5): 93–105 (in Chinese).

Bing, L. 2016. The strategies of integrating local knowledge into community-based disaster risk reduction: A case study of aba prefecture in Sichuan Province. *Journal of Guangzhou University (Social Science Edition)* 15 (6): 19–27 (in Chinese).

Bo, X. 2008. City disaster prevention and reduction plan research. TongJi University (in Chinese).

Bo, T. 2014. Analysis on urban disasters and countermeasures during urbanization. *Cities and Disaster Reduction* 5: 31–33 (in Chinese).

Bohong, W., and C. An. 2018. Framework of the evaluation model resilient cities. *Science & Technology Review* 36 (16): 94–99 (in Chinese).

Bozza, A., D. Asprone, and G. Manfredi. 2015. Developing an integrated framework to quantify resilience of urban systems against disasters. *Natural Hazards* 78 (3): 1729–1748.

Bruneau, M., S. E. Chang, R. T. Eguchi, et al. 2003. A framework to quantitatively assess and enhance the seismic resilience of communities. *Earthquake Spectra* 19 (4): 733–752.

Bruzzone, M., R. P. Dameri, and P. Demartini. 2021. Resilience reporting for sustainable development in cities. *Sustainability* 13 (14): 7824.

Campanella, T. J. 2006. Urban resilience and the recovery of New Orleans. *Journal of the American Planning Association* 72 (2): 141–146.

Carpenter, A. 2013. Social ties, space, and resilience. Community and Economic Development Discussion Paper, 1000. Community and Economic Development Department.

Carpenter, S., B. Walker, J. M. Anderies, et al. 2001. From metaphor to measurement: Resilience of what to what? *Ecosystems* 4 (8): 765–781.

Change, I. 2007. Climate change 2007: The physical science basis. *Agenda* 6 (07): 333.

Chelleri, L. 2012. From the «Resilient City» to urban resilience. A review essay on understanding and integrating the resilience perspective for urban systems. *Documents d'Anàlisi Geogràfica* 58 (2): 287–306.

Chong, C., and Y. Minxing. 2017. River basin governance based on resilient city theory. *Planners* 33 (8): 31–37.

Cutter, S. L., B. J. Boruff, and W. L. Shirley. 2012. *Social Vulnerability to Environmental Hazards*, 143–160. Routledge.

Cutter, S. L., K. D. Ash, and C. T. Emrich. 2014. The geographies of community disaster resilience. *Global Environmental Change* 29: 65–77.

Dell'Acqua, F., P. Gamba, and K. Jaiswal. 2013. Spatial aspects of building and population exposure data and their implications for global earthquake exposure modeling. *Natural Hazards* 68 (3): 1291–1309.

Desouza, K. C., and T. H. Flanery. 2013. Designing, planning, and managing resilient cities: A conceptual framework. *Cities* 35: 89–99.

Dongping, F., L. Zaishang, L. Nan, H. Linhai, et al. 2017. Urban resilience: A perspective of system of systems in trio spaces. *China Civil Engineering Journal* 50 (7): 1–7 (in Chinese).

Donsong, H., Z. Jian, and C. Zhan. 2014. The study of resilient community's building method and releasing path in the basis of intelligence technology. *Architecture & Culture* 7: 76–78 (in Chinese).

Enxin, G. 2016. Defensiveness, vulnerability and resilience: Three paradigms of urban safety studies. *Chinese Public Administration* 2016 (11): 105–110 (in Chinese).

Ernstson, H., S. E. Van Der Leeuw, C. L. Redman, et al. 2010. Urban transitions: On urban resilience and human-dominated ecosystems. *Ambio* 39 (8): 531–545.

Fangcheng, Y., and K. Hongjun. The "people-land" imbalance and its breakthrough in the new urbanization process. http://theory.people.com.cn/n1/2016/0801/c217905-28601100.html (in Chinese).

Fastiggi, M., S. Meerow, and T. R. Miller. 2021. Governing urban resilience: Organisational structures and coordination strategies in 20 North American city governments. *Urban Studies* 58 (6): 1262–1285.

Fengliang, X. 2012. The research of urban safety evaluation system construction. Hunan University of Science and Technology (in Chinese).

Folke, C. 2006. Resilience: The emergence of a perspective for social–ecological systems analyses. *Global Environmental Change* 16 (3): 253–267.

Folke, C., S. Carpenter, T. Elmqvist, et al. 2002. Resilience and sustainable development: Building adaptive capacity in a world of transformations. *AMBIO: A Journal of the Human Environment, BioOne* 31 (5): 437–440.

Folke, C., S. R. Carpenter, B. Walker, et al. 2010. Resilience thinking: Integrating resilience, adaptability and transformability. *Ecology and Society, JSTOR* 15 (4).

Frankenberger, T., and S. Nelson. 2013. Background paper for the expert consultation on resilience measurement for food security. *Expert Consultation on Resilience Measurement Related to Food Security*, 19–21.

Ge, L., N. P. Anten, I. D. van Dixhoorn, et al. 2016. Why we need resilience thinking to meet societal challenges in bio-based production systems. *Current Opinion in Environmental Sustainability* 23: 17–27.

Ge, L., Q. Purong, L. Xunchang, et al. 2020. Evaluation of city safety based on PSR level analysis model and BP neural network. *Journal of Catastrophology* 35 (3): 224–229 (in Chinese).

Godschalk, D. R. 2003. Urban hazard mitigation: Creating resilient cities. *Natural Hazards Review* 4 (3): 136–143.

Gordon, J. E. 2009. *Structures: Or Why Things Don't Fall Down*. Da Capo Press.

Guijuan, T. 2011. Research on integrated evaluation of urban natural disaster emergency capability. Harbin Institute of Technology (in Chinese).

Gunderson, L. H. 2000. Ecological resilience—In theory and application. *Annual Review of Ecology and Systematics* 31 (1): 425–439.

Gunderson, L. H. 2003. Adaptive dancing: Interactions between social resilience and ecological crises. In *Navigating Social-Ecological Systems: Building Resilience for Complexity and Change*, 33–52. Cambridge: Cambridge University Press.

Guofang, Z., and X. Chenhong. 2021. Strategic emphasis on the construction of resilient cities in China. *City Planning Review* 45 (2): 44–48 (in Chinese).

Guoping, Z., Z. Jiaqing, and F. Shengju. 2010. Local governments' response to natural disasters from the perspective of multi-bodies involved: Experience, problems and responses—A case of Yunnan Province. *Journal of Institute of Disaster Prevention* 12 (3): 120–124 (in Chinese).

Hallegatte, S., and P. Dumas. 2009. Can natural disasters have positive consequences? Investigating the role of embodied technical change. *Ecological Economics* 68 (3): 777–786.

Heng, Z., W. Haoyu, and C. Jiang. 2024. Research on the spatiotemporal evolution of urban resilience based on GA-BP neural network. *Journal of Shanxi Datong University (Natural Science Edition)* 40 (2): 71–77 (in Chinese).

Hengjing, W., F. Tienan, H. Zhonghua, et al. 2013. Earthquake casualty prediction model based on remote sensing images. *Journal of Tongji University (Medical Science)* 34 (5): 36–39 (in Chinese).

Holling, C. S. 1973. Resilience and stability of ecological systems. *Annual Review of Ecology and Systematics* 4 (1): 1–23.

Hong, W., L. Fengshan, and L. Zhipeng. 2018a. Topic review and future prospects of disaster management research in the 40 years of reform and opening up—A CiteSpace-based knowledge mapping analysis. *Guihai Tribune* 34 (5): 67–78, 2 (in Chinese).

Hong, H., L. Ruigi, F. Weicheng, and S. Chunchang. 2018b. Analysis on characteristics of safety resilient city and enlightenments for safe development of Xiongan New Area. *Journal of Safety Science and Technology* 14 (7): 5–11 (in Chinese).

Honghu, S., and Z. Feng. 2019. Evaluation of urban haze disaster resilience from the perspective of residents' activity: A case study of the main urban area of Nanjing city. *Scientia Geographica Sinica* 39 (5): 788–796 (in Chinese).

Hongrong, H. 2023. Resilience assessment of the key processes of grain crop-related physical-virtual water coupling flows in China. Northwest A&F University (in Chinese).

Hudec, O., A. Reggiani, and M. Šiserová. 2018. Resilience capacity and vulnerability: A joint analysis with reference to Slovak urban districts. *Cities* 73: 24–35.

Hui, Z., L. Yaolong, and F. Jieyao. 2020. Impact of urbanization quality and urban resilience on flood disaster risk: Based on panel data of 11 prefecture-level cities in Shanxi. *On Economic Problems* 4: 114–120 (in Chinese).

Huilin, H. How can we protect cultural heritage today? https://uci.sdu.edu.cn/info/1219/1311.htm (in Chinese).

Huiqing, W., X. Ruihua, L. Yanhui, et al. 2024. Visualization analysis of research literature on early warning of geo-hazards based on meteorological factors in the past 20 years. *The Chinese Journal of Geological Hazard and Control* 35 (2): 10–20 (in Chinese).
Index, C. R. 2014. City resilience framework, 928. The Rockefeller Foundation and ARUP.
Ishikawa, M. 2002. Landscape planning for a safe city. *Annals of Geophysics*.
Jabareen, Y. 2013. Planning the resilient city: Concepts and strategies for coping with climate change and environmental risk. *Cities* 31: 220–229.
Jha, A. K., T. W. Miner, and Z. Stanton-Geddes. 2013. *Building Urban Resilience: Principles, Tools, and Practice*. World Bank Publications.
Jianxin, Z. 2009. International comparison of integrated urban disaster prevention and mitigation planning. *Comparative Economic & Social Systems* 2: 171–174 (in Chinese).
Jianxin, Z., and L. Husheng. 2013. City public safety risk assessment-take high-rise building fire for example. *Journal of Safety Science and Technology* 9 (12): 133–137 (in Chinese).
Jieyao, F., L. Yaolong, W. Jun, and Z. Huaming. 2020. The impact of economic development and environmental pressure on urban resilience: Based on panel data of 11 prefecture-level cities in Shanxi. *Ecological Economy* 36 (9): 101–106, 163 (in Chinese).
Jinhe, Z., and S. Hongxue. 2020. Research on Spatial-temporal evolution and influencing factors of urban resilience of China's three metropolitan agglomerations. *Soft Science* 34 (2): 72–79 (in Chinese).
Joerin, J., R. Shaw, Y. Takeuchi, et al. 2014. The adoption of a climate disaster resilience index in Chennai, India. *Disasters* 38 (3): 540–561.
Jun, Y., X. Xiaohong, S. Qikun, et al. 2013. Prefectural- and county-level meteorological service platform based on 3D GIS. *Meteorological Science and Technology* 41 (1): 78–82 (in Chinese).
Junhui, J. 2017. Improvement of infrastructure to guard the peace of the community. *Disaster Reduction in China* 9: 28–29 (in Chinese).
Junlong, G., X. Hui, W. Chunyang, et al. 2024. Analysis on research progress of urban disaster prevention in China based on cite space. *City and Disaster Reduction* 1: 26–31 (in Chinese).
Kapucu, N., Y. Martín, and Z. Williamson. 2021. Urban resilience for building a sustainable and safe environment. *Urban Governance* 1 (1): 10–16.
Lam, N.S.-N., Y. Qiang, H. Arenas, et al. 2015. Mapping and assessing coastal resilience in the Caribbean region. *Cartography and Geographic Information Science* 42 (4): 315–322.
Lei, J. 2017. Pliable city design is the root of safe construction. *Shanghai Urban Management* 26 (3): 93–96 (in Chinese).
Lei, H., D. Shenzhi, and S. Yan. 2011. A research on multi-hazard mitigation planning in China with a reference to U.S. experience. *Urban Planning Forum* 5: 87–94 (in Chinese).
Leichenko, R. 2011. Climate change and urban resilience. *Current Opinion in Environmental Sustainability* 3 (3): 164–168.
Li, L. 2016. A brief analysis on the problems and measures of NGO's participation in public crisis management. In *2016 International Conference on Public Management*, 222–225. Atlantis Press.
Li, C., Z. Xigang, and S. Jie. 2017. The basic concept, mechanism and planning ideas of resilient cities. *Modern Urban Research* 9: 18–24 (in Chinese).
Li, Z., H. Shuang, L. Haibo, et al. 2023. Assessment and spatial-temporal characteristics of urban resilience in Beijing-Tianjin-Hebei region. *Project Management Technology* 21 (6): 89–95 (in Chinese).
Liang, Z. 2018. The concept of resilience in urban disaster prevention. *Beijing Planning Review* 2: 18–21 (in Chinese).
Liao, K.-H. 2012. A theory on urban resilience to floods—A basis for alternative planning practices. *Ecology and Society* 17 (4).
Limin, Z. 2016. Resilient city: Risk management and its indicators. *Journal of Beijing Administration Institute* 2: 13–20 (in Chinese).

Linjun, W., B. Yun, S. Shaolong, and W. Yuni. 2023. Knowledge structure and dynamic evolution of cross-border M&A study-based on a bibliometric analysis of Chinese core journal database from 1995 to 2019. *Management Review* 35 (9): 62–74 (in Chinese).
Liping, C., and Z. Fengqi. 2018. Enlightenment due to elastic construction in New York City for Shanghai. *Acta Ecologica Sinica* 38 (1): 86–95 (in Chinese).
Liu, D., Z. Fan, Q. Fu, et al. 2020. Random forest regression evaluation model of regional flood disaster resilience based on the whale optimization algorithm. *Journal of Cleaner Production* 250: 119468.
Liudan, J., D. Jiali, W. Ya, and H. Xiaosen. 2022. Evaluation of urban resilience based on PSR+cloud model. *Ecological Economy* 38 (5): 114–120 (in Chinese).
Longfei, L., S. Zhou, L. Zhe, et al. 2017. Disaster management system to support decision-making on disaster reduction and relief. *Disaster Reduction in China* 11: 18–21 (in Chinese).
Longyu, S., Z. Qiaoya, Y. Meng, and L. Lingyu. 2022. A review of definitions, influence factors and assessment of urban resilience. *Acta Ecologica Sinica* 42 (14): 6016–6029 (in Chinese).
Matos, F., V. M. Vairinhos, R. P. Dameri, et al. 2017. Increasing smart city competitiveness and sustainability through managing structural capital. *Journal of Intellectual Capital* 18 (3): 693–707.
Meerow, S., and J. P. Newell. 2017. Spatial planning for multifunctional green infrastructure: Growing resilience in Detroit. *Landscape and Urban Planning* 159: 62–75.
Meerow, S., J. P. Newell, and M. Stults. 2016. Defining urban resilience: A review. *Landscape and Urban Planning* 147: 38–49.
Meerow, S., P. Pajouhesh, and T. R. Miller. 2019. Social equity in urban resilience planning. *Local Environment* 24 (9): 793–808.
Mendizabal, M., O. Heidrich, E. Feliu, et al. 2018. Stimulating urban transition and transformation to achieve sustainable and resilient cities. *Renewable and Sustainable Energy Reviews* 94: 410–418.
Mileti, D. S. 1999. *By Design D. A Reassessment of Natural Hazards in the United State*. Washington (EUA): Joseph Henry Press of the National Academy of Sciences.
Min, L. 2021. Safety resilience evaluation of megacities in the context of natural disasters. *Project Management* 12: 68–71 (in Chinese).
Min, Z., Z. Wenbo, C. Xiaolin, et al. 2023. Research on resilience evaluation of sponge city. *Construction Economy* 44 (5): 5–12 (in Chinese).
Mingdou, Z., and F. Xiaoqing. 2018. Resilient cities: A new model for sustainable urban development. *Journal of Zhengzhou University (Philosophy and Social Sciences Edition)* 51 (2): 59–63 (in Chinese).
Minxing, Y., H. Bo, C. Chong, et al. 2016. Review and prospect of disaster prevention research based on resilient city theory. *Urban Planning Forum* 1: 48–55 (in Chinese).
Motesharrei, S., J. Rivas, E. Kalnay, et al. 2016. Modeling sustainability: Population, inequality, consumption, and bidirectional coupling of the earth and human systems. *National Science Review* 3 (4): 470–494.
Nan, S. 2017. Introductions to habitat III and new urban agenda, and their enlightenment on China. *City Planning Review* 1: 9–21 (in Chinese).
Ning, S. 2017. A study on measuring urban safety conditions in the Pearl River delta urban agglomeration. *Governance* 43: 33–48 (in Chinese).
Orencio, P. M., and M. Fujii. 2013. A localized disaster-resilience index to assess coastal communities based on an analytic hierarchy process (AHP). *International Journal of Disaster Risk Reduction* 3: 62–75.
Ouyang, M., L. Dueñas-Osorio, and X. Min. 2012. A three-stage resilience analysis framework for urban infrastructure systems. *Structural Safety* 36: 23–31.
Papa, R., A. Galderisi, M. C. V. Majello, et al. 2015. Smart and resilient cities. A systemic approach for developing cross-sectoral strategies in the face of climate change. *TeMA-Journal of Land Use, Mobility and Environment* 8 (1): 19–49.
Pelling, M. 2007. Learning from others: The scope and challenges for participatory disaster risk assessment. *Disasters* 31 (4): 373–385.

Peng, L. 2022. A comparative study on security governing of international smart city: A perspective of taking Tokyo, Singapore and Shanghai for example. Shanghai International Studies University (in Chinese).
Peng, C., Z. Lifeng, S. Yingyue, and L. Jiafu. 2015. The risk evaluation index system and the conceptual model of waterlogging disaster caused by excessive rain in urban areas. *Business & Economy* 11: 27–30 (in Chinese).
Pickett, S. T., M. L. Cadenasso, and J. M. Grove. 2004. Resilient cities: Meaning, models, and metaphor for integrating the ecological, socio-economic, and planning realms. *Landscape and Urban Planning* 69 (4): 369–384.
Pimm, S. L. 1984. The complexity and stability of ecosystems. *Nature* 307 (5949): 321–326.
Ponomarov, S. Y., and M. C. Holcomb. 2009. Understanding the concept of supply chain resilience. *The International Journal of Logistics Management* 20 (1): 124–143.
Qihui, X. 2017. Enlightenment of resilient city construction policy in developed country. *Scientific Decision Making* 4: 60–75 (in Chinese).
Qin, Z., and G. Jiacheng. 2023. Practical exploration, challenges and realistic paths of modernizing resilient governance in Chinese-style urban communities. *Social Governance Review* 5: 100–107 (in Chinese).
Qin, Z., J. Yuanyuan, and J. Jun. 2010. Exploring the coupling mechanism of social organization participation in public crisis governance. *Theoretical Investigation* 2: 131–135 (in Chinese).
Reghezza, M., and S. Rufat. 2015. *Resilience Imperative. Uncertainty, Risks and Disasters.* Elsevier.
Resilient cities, thriving cities: The evolution of urban resilience. https://iclei.org/e-library/resilient-cities-thriving-cities-the-evolution-of-urban-resilience-11/
Ribeiro, P. J. G., and L. A. P. J. Gonçalves. 2019. Urban resilience: A conceptual framework. *Sustainable Cities and Society* 50: 101625.
Ruidong, Z., F. Chuanglin, and L. Haimeng. 2020. Progress and prospect of urban resilience research. *Progress in Geography* 39 (10): 1717–1731 (in Chinese).
Sendai framework for disaster risk reduction 2015–2030.
Several thoughts on cultivating safety culture in the New Era. file:///C:/Users/Ling/Zotero/storage/ZD4JN7P7/t20210202_2251256.html (in Chinese).
Sharifi, A., and A. R. Khavarian-Garmsir. 2020. The COVID-19 pandemic: Impacts on cities and major lessons for urban planning, design, and management. *Science of the Total Environment* 749: 142391.
Shengdong, L., S. Junnan, C. Yong, et al. 2020. Review of pollution characteristics of PM2.5 in Chinese representative megacities. *Research of Environmental Sciences* 33 (2): 243–251 (in Chinese).
Shi, Y., G. Zhai, L. Xu, et al. 2021. Assessment methods of urban system resilience: From the perspective of complex adaptive system theory. *Cities* 112: 103141.
Sisi, Z., M. Xiaoyu, X. Jiangling, and C. Qi. 2023. The impact of population agglomeration on urban resilience. *Northwest Population Journal* 44 (1): 76–90 (in Chinese).
Smit, B., and J. Wandel. 2006. Adaptation, adaptive capacity and vulnerability. *Global Environmental Change* 16 (3): 282–292.
Spaans, M., and B. Waterhout. 2017. Building up resilience in cities worldwide—Rotterdam as participant in the 100 resilient cities programme. *Cities* 61: 109–116.
Tahmasebi Birgani, Y., and F. Yazdandoost. 2018. An integrated framework to evaluate resilient-sustainable urban drainage management plans using a combined-adaptive MCDM technique. *Water Resources Management* 32 (8): 2817–2835.
The '8-4' mega explosion at the Port of Beirut, Lebanon in 2020. https://news.sina.com.cn/o/2020-12-29/doc-iiznezxs9580721.shtml (in Chinese).
The collapse of Xinjia Hotel in Quanzhou killed 29 people. https://www.thepaper.cn/newsDetail_forward_10923560 (in Chinese).
The investigation report of Jiangsu Xiangshui Tianjiayi Chemical Co., Ltd. on the "March 21" extraordinarily serious explosion accident was announced. https://www.gov.cn/xinwen/2019-11/15/content_5452468.htm (in Chinese).

Thomas, T., L. Christiaensen, Q.-T. Do, et al. 2010. Natural disasters and household welfare: Evidence from Vietnam. World Bank Policy Research Working Paper, 5491.

Ting, H., and D. Zaipeng. 2017. Study on index system of situation safety assessment for urban complex. *Guangzhou Chemical Industry* 45 (4): 95–97 (in Chinese).

Tyler, S., and M. Moench. 2012. A framework for urban climate resilience. *Climate and Development* 4 (4): 311–326.

van den Berg, L. 2006. *The Safe City: Safety and Urban Development in European Cities*. Ashgate Publishing, Ltd.

Walker, B. H., N. Abel, J. M. Anderies, et al. 2009. Resilience, adaptability, and transformability in the Goulburn-Broken Catchment, Australia. *Ecology and Society* 14 (1).

Wang, L., Z. Gong, L. Shi, et al. 2021. Knowledge mapping analysis of research progress and frontiers in integrated disaster risk management in a changing climate. *Natural Hazards* 107 (3): 2033–2052.

Wei, D. 2023. Study on the planning and design of Pearl River delta based on the thought of resilience. South China University of Technology (in Chinese).

Weicheng, F. 2020. Trends in the development of safe and resilient cities. *Labour Protection* 3: 20–23 (in Chinese).

Wekerle, G. R., and C. Whitzman. 1995. *Safe Cities: Guidelines for Planning, Design and Management*. USA: A Division of International Thomson Publishing Inc.

Wenfang, C., X. Wei, and S. Peijun. 2011. Risk assessment of typhoon disaster at county level in the Yangtze River Delta of China. *Journal of Natural Disasters* 20 (4): 77–83 (in Chinese).

Wenqian, D. 2020. Urban comprehensive disaster risk assessment. North China University of Science and Technology (in Chinese).

Wenya, W., W. Mingfeng, Z. Yinghao, and W. Fan. 2023. Research progress of internet and digital rural in China. *World Regional Studies* 32 (2): 70–81 (in Chinese).

Xiaoguang, W., and L. Ying. 2022. Urban economic vitality: Characteristics, evaluation system and promotion proposals. *Regional Economic Review* 1: 130–138 (in Chinese).

Xiaojun, H., and H. Xin. 2015. Resilient city and its planning framework. *City Planning Review* 39 (2): 50–56 (in Chinese).

Xiaoli, L., and G. Wanshan. 2007. Comprehensive evaluation index system for urban economic vitality. *Statistics & Decision* 11: 77–78 (in Chinese).

Xiaoliang, Q., L. Xuetao, and L. Ruxia. 2019. Problems and countermeasures in governmental natural disaster emergency management. *Journal of Liaoning Academy of Governance* 4: 61–65 (in Chinese).

Xiaoya, O. 2012. Comprehensive urban disaster risk assessment. Jiangxi University of Technology (in Chinese).

Xinghua, F. 2023. Study on the safety of spatial development of Shenyang from the perspective of landscape patterns. Northeast Normal University (in Chinese).

Xinyu, Z., and W. Qiao. 2019. The evolution of the urban resilience concept, and its research contents and development trend. *Science & Technology Review* 37 (22): 94–104 (in Chinese).

Xixi, S., and Y. Li. 2020. Application of ecological security evaluation model of Chengdu-Chongqing urban agglomeration. *Journal of Heilongjiang University of Technology (Comprehensive Edition)* 20 (11): 93–103 (in Chinese).

Xue, L., Y. Hongxia, and L. Peng. 2018. Resilience-based seismic performance concept, assessment methodology and engineering application. *Building Structure* 48 (18): 1–7 (in Chinese).

Yali, Z. 2007. The crisis of public safety management in China. *Journal of the Party School of CPC Zhengzhou Municipal Committee* 3: 48–49 (in Chinese).

Yan, Z. 2013. Promote urban adaptation to the planning, build cities with tenacity and toughness. *World Environment* 6: 50–53 (in Chinese).

Yanfen, H., H. Xintong, Y. Zekun, et al. 2018. Study on the UK disaster prevention, mitigation and relief system (above). *Disaster Reduction in China* 21: 58–61 (in Chinese).

Yaru, C. 2023. Study on the evaluation of urban resilience in the post-epidemic period—Taking the Yangtze River delta urban agglomeration as an example. Shandong Jianzhu University (in Chinese).

Yibao, W. 2018. Urban development in the context of a larger security perspective. *The Masses* 4: 11–12 (in Chinese).

Ying, L., and Z. Wangfeng. 2022. Research progress of urban resilience in China. *Journal of Green Science and Technology* 24 (21): 235–241, 246 (in Chinese).

Yiwei, C. 2023. Industrial agglomeration and urban economic resilience. Southeast University (in Chinese).

Yiwen, S., and X. Jiang. 2015. Understanding urban resilience: A conceptual analysis based on integrated international literature review. *Urban Planning International* 30 (2): 48–54 (in Chinese).

Yongxiang, L. 2011. What is disaster?—Identifying core concepts in the anthropological study of disasters. *Journal of Southwest Minzu University (Humanities and Social Sciences Edition)* 32 (11): 12–20 (in Chinese).

Yuanbing, P., Z. Xi, and S. Chaofeng. 2023. Study on the impact of ATIS on the resilience of urban road traffic network after a disaster. *Journal of Chongqing University of Technology (Natural Science)* 37 (3): 39–46 (in Chinese).

Yudong, Z. 2021. Research on risk diagnosis and adaptive mechanism of urban areas oriented to fire control and safety. Capital University of Economics and Business (in Chinese).

Yujie, Z., and W. Xunqing. 2023. CiteSpace-based urban resilience research history and hot spot analysis. *Future City Design & Management* 1: 19–23 (in Chinese).

Yun, L., P. Jingjing, et al. 2005. Design of safety indicator system for a city of well-off society. *China Safety Science Journal* 1: 27–31, 21, 2–3 (in Chinese).

Zezhao, L., and G. Runnan. 2023. Comparative analysis for urban resilience research among China and foreign countries. *Journal of Beijing City University* 5: 26–35 (in Chinese).

Zhang, H., J. Yang, L. Li, et al. 2021a. Measuring the resilience to floods: A comparative analysis of key flood control cities in China. *International Journal of Disaster Risk Reduction* 59: 102248.

Zhang, J., M. Zhang, and G. Li. 2021b. Multi-stage composition of urban resilience and the influence of pre-disaster urban functionality on urban resilience. *Natural Hazards* 107 (1): 447–473.

Zhang, L., Y. Wang, J. Zhang, et al. 2022. Rockfall hazard assessment of the slope of Mogao Grottoes, China based on AHP, F-AHP and AHP-TOPSIS. *Environmental Earth Sciences* 81 (14): 377.

Zhaofeng, X., T. Jiefang, and Z. Jing. 2019. Urban resilience evaluation system and optimization strategy from perspective of disaster prevention. *China Safety Science Journal* 29 (3): 1–7 (in Chinese).

Zhen, Z., W. Zeyu, and L. Ying. 2021. Review of researches on the high quality development of regional economy in China. *Resource Development & Market* 37 (8): 928–933 (in Chinese).

Chapter 2
Evaluation Index System for Urban Safety Resilience Coupling Coordination

It is crucial to consider the complexity, phasic nature, and dynamism of urban safety resilience across regions to develop an evaluative framework for coordinating urban safety resilience. The framework should adhere to scientific validity, system efficiency, and quantitative objectivity principles, mainly focusing on disaster prevention and reduction. The principles ensure that the indices are based on scientific principles, are efficient in computation, and can be measured objectively. They encompass various domains such as welfare, health, society, economy, environment, infrastructure, leadership, and planning. All of these areas are critical when mitigating the impacts of disasters. This approach provides a comprehensive metric for assessing urban safety resilience, especially during natural disasters and other crises. The evaluative framework prioritizes urban personnel safety resilience, urban facility safety resilience, and urban management safety resilience, which are essential for effective disaster response. The system is organized into multiple standard levels, which are further separated into sub-standards and specific indicators related to disaster readiness and response capabilities. This detailed structure allows for an in-depth assessment of various aspects of urban safety and resilience, facilitating accurate and complete evaluations, thereby effectively enhancing strategies for urban safety resilience in the face of disasters.

Main Principles for Evaluation Index System

Urban safety resilience represents a complex system, with the level of urban safety resilience in different regions exhibiting comprehensive, phased, and dynamic characteristics. It is impossible to completely and accurately gauge the coupling coordination of urban safety resilience among different regions using a single index. It calls for adopting a multi-index comprehensive study and assessment technique to develop a suitable index system. The evaluation approach divides the research topic into

Y. Xiang et al., *Research on Coupling Coordination of Urban Safety Resilience Based on Disaster Prevention and Reduction*, Resilient and Friendly Living Environments,
https://doi.org/10.1007/978-981-96-1562-9_2

different levels, selecting several independent indices and integrating them according to a hierarchical structure. Subsequently, appropriate weights are assigned to each index through subjective or objective methods to establish the model and complete the objective evaluation. The selection of indices for the urban safety resilience system and the construction of evaluation models must adhere to the following basic principles:

(1) Principle of Scientificity and Rationality

To create a measurement system for coordinating urban safety resilience, adhering to the principles of sustainable development, green development, and coupling coordination theory is necessary. The selected indices and model designs must conform to these objectives while accurately and impartially capturing the current situation and trends in urban safety and urban resilience in different geographical areas. When choosing indices, it is essential to make sure that the selected indices provide a reasonable interpretation of the goal theme, originate from accurate and authoritative sources, and maintain a certain level of independence to prevent duplication (Hui 2018). The selection of indices should be based on scientific reasoning, striking a balance between simplicity of calculations and comprehensive representation of all aspects of the urban safety resilience system (Ran 2017). It will simplify the index system while ensuring accurate calculation results for coordinating urban safety resilience.

(2) Principle of Systematicness and Simplification

Assessing the coordination of urban safety resilience is a comprehensive study that analyzes several aspects of a complex system. When developing the index system, employing systematic logic and thoroughly assessing and quantifying criteria is imperative (Hongyan 2019). Factors such as urban personnel safety, infrastructural safety, and emergency management safety should be considered. Systematics dictates that the choice of indices should not be comprehensive but rather adhere to Occam's Razor. It means selecting crucial indices that emphasize the features of urban safety resilience in order to simplify the system and decrease computing complexity.

(3) Principle of Quantifiability and Objectivity

Urban safety resilience coupling coordination needs to integrate theoretical frameworks and empirical facts. The lack of sufficient data poses a challenge in conducting research and establishing a reference point for enhancing urban safety and resilience. The evaluation indices selected must be measurable and feasible, focusing on quantitative indices derived from reliable, standardized data to guarantee the impartiality and logicality of the analysis. When determining the weights of indices, objective weighing procedures are recommended to minimize the impact of human and subjective factors (Ruihua 2018).

Preliminary for Constructing the Evaluation Index System

Selection of Evaluation Indices

Selecting suitable evaluation indices is the first step in developing a coupling coordination index system. A robust evaluation index system is required to ensure accurate results. Fan Weicheng suggested that a city consists of three primary components: individuals, items, and operational systems. He also introduced the safety-resilience triangle model (Weicheng 2020), shown in Fig. 2.1. The model entails conducting research on events, carrier systems, and safety resilience management, emphasizing the importance of comprehensive coordination and implementation planning to enhance urban resilience.

This book refers to the *Triangle Theory of Safety and Resilience* proposed by Academician Fan and *Guide for Safety Resilience Evaluation* (GB/T 40947-2021) proposed by the Chinese national standard (Wan et al. 2022). This book focuses on the constituent components of resilience in urban construction to expand the theories and the coupling mechanism of urban safety resilience. The approach follows globally acknowledged academic universities and institutions (e.g., Rockefeller Foundation and Hosei University). These standards cover many areas, such as well-being and

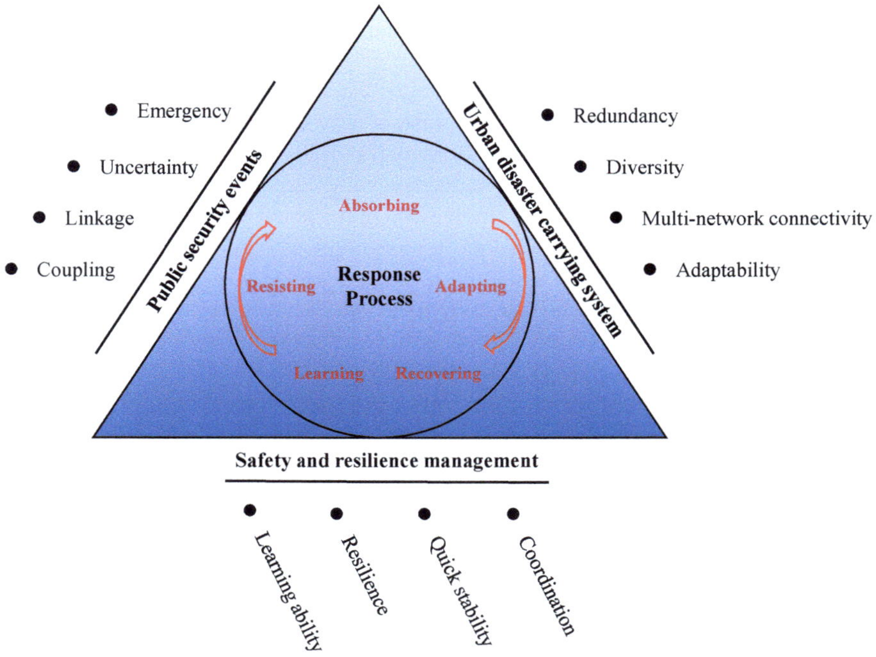

Fig. 2.1 Triangular theory of security and resilience (Weicheng 2020)

health, society and the economy, environment and infrastructure, and leadership and planning. The theme has been carefully evaluated by considering urban and administrative factors. The process has resulted in selecting indices that most effectively align with the research objective, which establishes a rigorous and precise assessment index system.

1. *Guide for Safety Resilience Evaluation* (GB/T 40947-2021)

Chinese authorities have collaborated with the large R&D project "Construction of Safety Resilience Cities and Disaster Prevention Technology Research and Demonstration" as part of the 13th Five-Year Plan. The collaboration aims to strengthen the construction of resilient urban environments and assure stable urban operations. The *Guide for Safety Resilience Evaluation* was officially proposed in November 26th, 2021. The national standard is crucial for normalizing, including, and standardizing evaluations related to the resilience of urban safety. This book utilizes the public safety triangle theory and the four elements of safety management to analyze urban safety resilience. It evaluates the coordination and evaluation of personnel, facility, and management.

2. Sustainable Development Goals

The United Nations proposed the Millennium Development Goals on September 25th, 2000, which led to the formation of sustainable development goals (SDGs). All 193 member countries attended a summit that the United Nations convened at its headquarters in New York in September 2015. The conference officially approved the document titled "Transforming our World: The 2030 Agenda for Sustainable Development," which outlined 17 objectives and 169 specific targets for achieving sustainable development. Under the direction of the 2030 Agenda, the United Nations Statistics Commission established the Inter-agency Expert Group on SDG Indicators (IAEG-SDGs) to efficiently implement these goals and track the advancement of sustainable development in several domains. The goal of this group is to create a flexible framework for assessing indicators and improving the general and specific objectives of sustainable development, which are currently being revised and enhanced. The newest *Global Sustainable Development Progress Report* highlights the formidable challenge of achieving the goals of the 2030 Agenda in light of current development trends. Even highly developed countries and regions struggle to satisfy particular standards. China and other developing nations face the pressing need for economic transformation and industrial upgrading. Attaining the SDGs would require a lengthy process for these countries. In addition, due to global warming and numerous human activities, severe natural disasters are increasing, which presents a substantial obstacle to national efforts in disaster prevention and mitigation management. Hence, it is essential to use the SDGs to advance the development of urban safety resilience in the context of disaster prevention and mitigation. Figure 2.2 illustrates the United Nations' sustainable development structure.

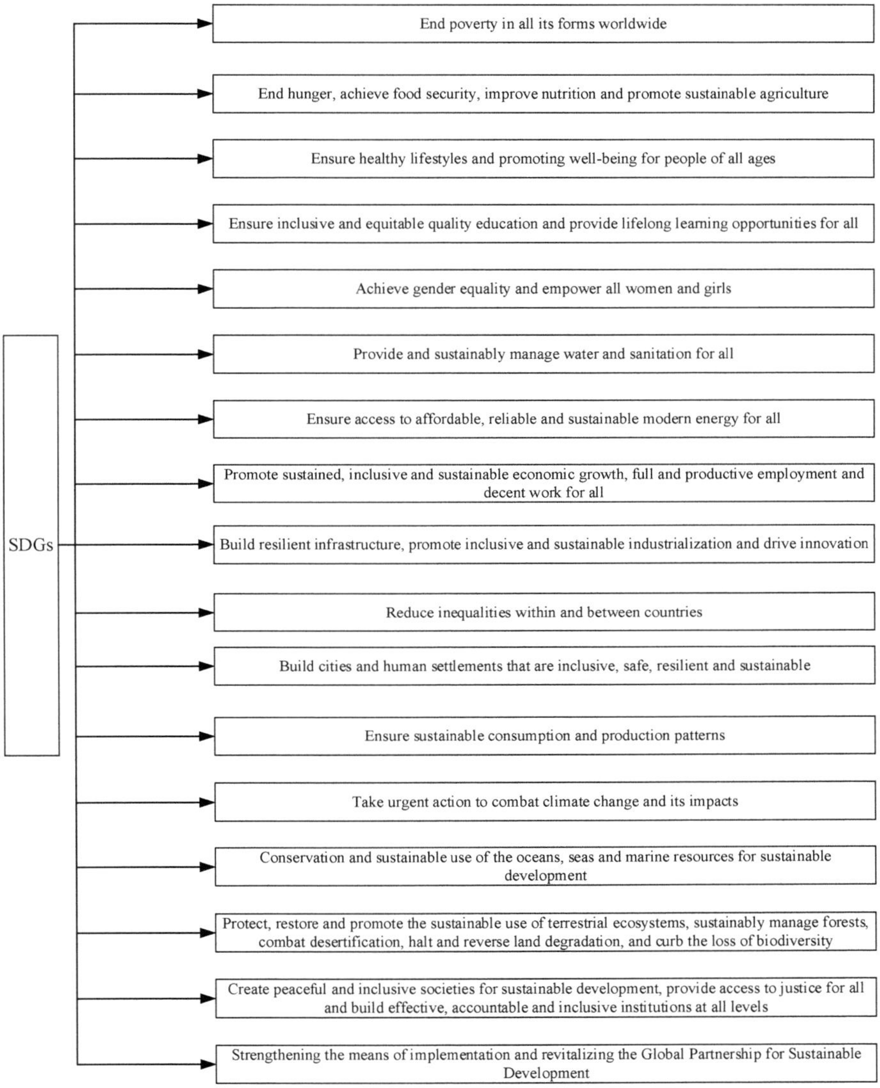

Fig. 2.2 United Nations overall framework for sustainable development

Given these challenges, this book emphasizes improving the cleanliness of urban environments, constructing infrastructure that can withstand disasters, and encouraging healthy habits among city dwellers, all within the context of the SDGs. While conflicts may exist between the goals of safety resilient cities and the SDGs, examining their interconnections and the possibility of mutually beneficial coexistence is essential.

3. Preliminary Framework of the Urban Safety Resilience Coupling Coordination Evaluation Index System

The urban safety resilience coupling coordination evaluation index system is built upon the *Guide for Safety Resilience Evaluation* (GB/T 40947-2021) and considers the SDGs. It consists of objective, criterion, sub-criterion, and index levels. We can thoroughly assess and enhance urban safety resilience by employing this systematic method. The objective level comprises multiple sub-criterion levels that concentrate on fundamental demographic characteristics, social participation preparedness, and a culture of safety and security for assessing urban personnel safety resilience. These sub-criterion levels include urban personnel safety resilience, urban facility safety resilience, and urban management safety resilience. Sub-criterion levels for urban facility safety resilience include construction engineering, transportation facilities, lifeline engineering facilities, and emergency support facilities, each with specific indices. For urban management safety resilience, the sub-criterion levels include risk control level, support and investment, and material supply level, each detailed with specific indices to assess the overall situation in these areas.

(1) Urban Personnel Safety Resilience

Urban personnel safety resilience consists of three sub-criterion: basic population attributes, social participation readiness, and safety perception and safety culture. There are a total of 14 distinct indications. These indicators include the density of permanent residents in the built-up area, the level of basic medical insurance for urban workers, the proportion of temporary residents, and the proportion of the disabled population. These indicators are utilized to assess the overall size of the population in the region. The social community's readiness is determined by three indicators: the quantity of skilled health staff available in the city, the level of preparedness of urban health institutions, and the capacity of social organization units. These indicators indicate the level of social catastrophe response in the region. The indicators for safety perception and safety culture consist of four factors: the extent of coverage under unemployment insurance, urban old-age insurance, work-related injury insurance, and the per capita emergency deposit level. These indicators offer an overview of the region's population, property, and social security condition.

(2) Urban Facility Safety Resilience

Urban facility safety resilience includes four sub-criterion: construction engineering, transportation facilities, lifeline engineering facilities, and emergency support facilities, with 13 specific indicators. Construction engineering involves four indicators: urbanization rate, the proportion of land area in safety-vulnerable areas, the proportion of earthquake-resistant buildings, and the level of personnel reserves in the construction industry. These indicators are used to assess the disaster prevention and mitigation development level in the region's construction industry. Transportation facilities involve two indicators: the per capita urban road area and the level of urban traffic lighting facilities. These indicators reflect the level of transportation facilities in the region. Lifeline engineering facilities involve two indicators: the penetration rate of wired broadcasting and television and the level of urban drainage facilities. These indicators are used to portray the overall situation of lifeline engineering facilities in the region. Emergency support facilities involve five indicators: per capita refuge area, public coverage of meteorological disaster monitoring, forecasting, and warning information, building area of disaster relief reserve institution warehouses per ten thousand people, the number of beds in medical and health institutions per ten thousand people, and the green coverage rate. These indicators are employed to gauge the level of development of emergency support facilities in the region.

(3) Urban Management Safety Resilience

Urban management safety resilience includes three sub-criterion: risk control level, support and guarantee input level, and material supply level, with eight specific indicators. The risk control level involves two indicators: the death rate per million people due to disasters and the proportion of annual direct economic losses due to disasters to the region's gross domestic product. These indicators assess the overall risk control level in the region. The support and guarantee input level involves three indicators: the level of financial support for public safety, the level of financial support for medical and health care, and the level of financial support for transportation. These indicators reflect the level of support and input in urban safety management. The material supply level involves three indicators: the comprehensive production capacity of water supply, grain supply, and emergency material supply. These indicators depict the overall material supply situation in the region (Ya et al. 2016).

Initial Indicator Description

The detailed description of each indicator and data processing method is shown in Table 2.1. The indicator types in the table are positive and negative. Positive indicators denote variables with higher values indicating better outcomes, whereas negative indicators denote variables with lower values indicating better outcomes.

Optimization of the Initial Evaluation Index System

1. Optimization of the Index System Through Expert Consultation Method

The original collection of evaluation indices, primarily derived from relevant research, may contain features that are not realistic. Continuously develop and modify the index system to enhance its scientific validity. To achieve this goal, we conducted a consultation using expert questionnaires to assess the first collection of 35 indices that measure the coupling coordination of urban safety resilience. We requested professionals with expertise to delete, alter, add, or reclassify these indexes, systematically improving the system through their experience, knowledge, and judgment.

Table 2.1 Framework of urban safety resilience coupling coordination evaluation index system

Target layer	Criterion layer	Sub-criterion layer	Indicator layer	Unit	Indicator type
Urban safety resilience coupled coordinated development index	Urban personnel safety resilience	Basic population attributes	The density of permanent residents in the built-up area	Ten thousand people/ square kilometer	Positive
			The level of basic medical insurance for urban workers	Ten thousand people	Positive
			The proportion of temporary residents	%	Negative
			The proportion of disabled population	%	Positive
		Social participation readiness	The level of reserve health technical personnel in the city	Individual	Positive
			The reserve level of urban health institutions	Individual	Positive
			The number of police officers in ten thousand	Individual	Positive
			The number of firefighters in ten thousand	Individual	Positive
			The level of social organization units	Individual	Positive
		Safety perception and safety culture	The number of people covered by unemployment insurance	Ten thousand people	Positive
			The number of people covered by urban old-age insurance	Ten thousand people	Positive
			The number of people covered by work-related injury insurance	Ten thousand people	Positive
			Commercial insurance density	%	Positive
			The per capita emergency deposit level	Yuan/person	Positive

(continued)

Table 2.1 (continued)

Target layer	Criterion layer	Sub-criterion layer	Indicator layer	Unit	Indicator type
	Urban facility safety resilience	Construction engineering	Urbanization rate	%	Positive
			Proportion of land area in safety-vulnerable areas	%	Negative
			Proportion of earthquake-resistant buildings	%	Positive
			The level of personnel reserves in the construction industry	Ten thousand people	Positive
		Transportation facilities	Per capita urban road area	Square meters	Positive
			The level of urban traffic lighting facilities	Lamp	Positive
		Lifeline engineering facilities	The penetration rate of wired broadcasting and television	%	Positive
			The level of urban drainage facilities	Kilometer	Positive
		Emergency support facilities	The per capita refuge area	Square meters/ person	Positive
			Public coverage of meteorological disaster monitoring, forecasting and warning information	%	Positive
			Building area of disaster relief reserve institution warehouses per ten thousand people	Square meters/ten thousand people	Positive
			The number of beds in medical and health institutions per ten thousand people	Individuals/ ten thousand people	Positive
			The green coverage rate	%	Positive
	Urban management safety resilience	Risk control level	The death rate per million people due to disasters	%	Negative
			The proportion of annual direct economic losses due to disasters to the region's gross domestic product	%	Negative

(continued)

Table 2.1 (continued)

Target layer	Criterion layer	Sub-criterion layer	Indicator layer	Unit	Indicator type
		Support and guarantee input level	The level of financial support for public safety	100 million yuan	Negative
			The level of financial support for medical and health care	100 million yuan	Positive
			The level of financial support for transportation	100 million yuan	Positive
		Material supply level	The comprehensive production capacity of water supply	Ten thousand cubic meters/day	Positive
			The level of grain supply	Ten thousand tons	Positive
			The level of emergency material supply	100 million yuan	Positive

The expert consultation primarily focused on several crucial aspects: the effectiveness of the dimension setup, the presence of overlap or redundancy among the indices, the placement of the indices within the evaluation dimensions, the accuracy and objectivity of the indices in reflecting real-world urban safety resilience, and the ease of using the indices in the evaluation process. Appendix A provides a detailed description of the format and organization of the expert opinion request.

The expert consultation approach, which mainly depends on subjective judgments, is vulnerable to variations in expert perceptions and professional expertise, which can impact the precision of the evaluation outcomes. The selection and consultation process followed these basic principles: Emergency management professionals are required to have at least five years of practical or instructional experience. Experts from administrative departments, higher education/scientific institutions, corporate sectors, and professional groups were recruited to guarantee comprehensive and reliable survey findings. A group of 20 emergency management experts has been established to reduce the influence of personal prejudices on the accuracy of the evaluation findings. The consultation process is independently performed, including face-to-face interviews. During the interviews, professionals must be required to fill out an index system expert screening form and a completeness opinion solicitation form. An index was retained if it received more than two-thirds of the total number of experts; it was removed if it received less than two-thirds.

We compiled and categorized the results of this expert consultation, offering the following recommendations:

(1) Removing "the proportion of the disabled population" from the sub-criterion of basic demographic attributes. Although it is an important social indicator,

it should be considered separately in city planning and emergency response strategies rather than being directly utilized as an evaluation index.

(2) Adding "the number of newly added higher education talents" to the sub-criterion of basic demographic attributes. Increasing educated talents can enhance a city's social stability and crisis response capabilities.
(3) Removing "the number of police officers in ten thousand" and "the number of firefighters in ten thousand" indices from the sub-criterion of societal participation readiness due to the difficulty of obtaining this data, which is frequently classified.
(4) Removing "commercial insurance density" from the sub-criterion layer of safety perception and safety culture, as it partly overlaps with "the number of people covered by urban old-age insurance" and "the number of people covered by unemployment insurance." Policy, market, and industry changes can all have an impact.
(5) Suggesting the deletion of "proportion of earthquake-resistant buildings" and "public coverage rate of meteorological disaster forecasting and warning information" from the criterion of architectural engineering and emergency preparedness facilities, respectively. The reason for this is the existing difficulty in acquiring extensive macro data for these indices.

After conducting these consultations, the revised Urban Safety Resilience Coupling Coordination Evaluation Index System has been updated to incorporate three criteria, ten sub-criterion, and thirty specific indicators, as displayed in Table 2.2. This modified system aims to provide a more holistic and accurate measure of a city's capacity to manage and mitigate risks associated with safety and resilience.

2. Validity and Reliability Analysis of the Index System

Due to the scarcity of professionals in the first step of index optimization, it is imperative to ensure that the constructed index system comprehensively reflects the safety resilience of cities. The second step involves expanding the number of participants and employing a 10-point Likert Scale to optimize the set of innovation indices for the construction industry. The Likert scale is commonly employed for its ability to evaluate closed-ended questions and enables the assessment of the importance of indices, as seen in Appendix B.

The online survey using the Questionnaire Star website was conducted. Two primary criteria was introduced to ensure the survey results were correct: First, respondents were required to possess a minimum of two years of experience in emergency management, social work, or academic research. Second, the experts from government departments, businesses, professional organizations, and higher education or scientific institutions guarantee the comprehensiveness and accuracy of the results. A sample of 100 individuals employed in emergency management was surveyed to mitigate the influence of personal biases on the accuracy of the results.

A two-step examination was conducted to improve the accuracy of the survey data: First, a thorough review of the results to exclude any responses that exhibited a missing rate over 10%, had identical answers, or displayed an obvious pattern.

Table 2.2 Urban safety resilience coupling coordination evaluation index system

Target layer	Criterion layer	Sub-criterion layer	Indicator layer	Unit	Indicator description	Indicator type
Urban safety resilience coupled coordinated development index	Urban personnel safety resilience	Basic population attributes	The density of permanent residents in the built-up area	Ten thousand people/square kilometer	Refers to the ratio of the number of permanent residents in the built-up area to the area of the built-up area; the calculation method is: number of permanent residents in the built-up area/area of the built-up area	Positive
			The level of basic medical insurance for urban workers	Ten thousand people	Refers to the number of people who participate in the corresponding urban employee basic medical insurance in accordance with relevant national regulations	Positive
			The proportion of temporary residents	%	Refers to the percentage of urban temporary resident population in the total urban population; the calculation method is: urban temporary resident population/(total urban permanent resident population + urban temporary resident population) × 100%	Negative
			The number of newly added higher education talents	Ten thousand people	Refers to the number of graduates from Jintong undergraduate (college) schools in each city (state)	Positive
		Social participation readiness	The level of reserve health technical personnel in the city	Individual	Refers to the number of health technicians per ten thousand permanent residents in a city	Positive
			The reserve level of urban health institutions	Individual	Refers to the number of health institutions in each city (state)	Positive
			The level of social organization units	Individual	Refers to the number of social organizational units in a city within a certain period of time	Positive

(continued)

Table 2.2 (continued)

Target layer	Criterion layer	Sub-criterion layer	Indicator layer	Unit	Indicator description	Indicator type
		Safety perception and safety culture	The number of people covered by unemployment insurance	Ten thousand people	Refers to the number of urban unemployment insurance insured persons at the end of a certain period of time	Positive
			The number of people covered by urban old-age insurance	Ten thousand people	Refers to the number of urban pension insurance insured persons at the end of a certain period of time	Positive
			The number of people covered by work-related injury insurance	Ten thousand people	Refers to the number of people insured by work-related injury insurance at the end of a certain period of time	Positive
			The per capita emergency deposit level	Yuan/person	Refers to the ratio of urban household deposits to the total urban resident population; the calculation method is: urban household deposits/total urban resident population	Positive
	Urban facility safety resilience	Construction engineering	Urbanization rate	%	Refers to the ratio of urban population to total population in a region	Positive
			Proportion of land area in safety-vulnerable areas	%	Refers to the percentage of land area in security-vulnerable areas to construction land area in built-up areas; the calculation method is: land area in security-vulnerable areas/construction land area in built-up areas × 100%	Negative
			The level of personnel reserves in the construction industry	Ten thousand people	Refers to the number of employees in construction enterprises in a city within a certain period of time	Positive

(continued)

Table 2.2 (continued)

Target layer	Criterion layer	Sub-criterion layer	Indicator layer	Unit	Indicator description	Indicator type
		Transportation facilities	Per capita urban road area	Square meters	Refers to the average urban road area per person in the urban area	Positive
			The level of urban traffic lighting facilities	Lamp	Refers to the number of road lighting lights in a city within a certain period of time	Positive
		Lifeline engineering facilities	The penetration rate of wired broadcasting and television	%	Refers to the actual users of cable radio and television in the total population of the administrative region/the total number of permanent residents in the city	Positive
			The level of urban drainage facilities	Kilometer	Refers to the length of urban drainage pipes within a certain period of time	Positive
		Emergency support facilities	Per capita refuge area	Square meters/person	Refers to the ratio of the area of urban emergency shelters to the total urban resident population; the calculation method is: area of urban emergency shelters (square meters)/total urban resident population (persons)	Positive
			Building area of disaster relief reserve institution warehouses per ten thousand people	Square meters/ten thousand people	Refers to the ratio of the emergency supplies reserve area to the total urban resident population (ten thousand people); the calculation method is: emergency supplies reserve area (square meters)/total urban resident population (ten thousand people)	Positive

(continued)

Table 2.2 (continued)

Target layer	Criterion layer	Sub-criterion layer	Indicator layer	Unit	Indicator description	Indicator type
			The number of beds in medical and health institutions per ten thousand people	Individuals/ten thousand people	Refers to the ratio of the number of beds in various medical and health institutions to the total urban resident population (ten thousand people); the calculation method is: the number of beds in various medical and health institutions/total urban resident population	Positive
			The green coverage rate	%	Refers to the percentage of the total green coverage area in the urban built-up area to the total area of the urban built-up area	Positive
	Urban management safety resilience	Risk control level	The death rate per million people due to disasters	%	Refers to the percentage of annual disaster-related deaths to the total urban resident population (millions); the calculation method is: annual disaster-related deaths/total urban resident population (millions) × 100%	Negative
			The proportion of annual direct economic losses due to disasters to the region's gross domestic product	%	Refers to the proportion of annual direct economic losses due to disasters to regional GDP; the calculation method is: annual direct economic losses due to disasters/regional GDP × 100%	Negative
		Support and guarantee input level	The level of financial support for public safety	100 million yuan	Refers to urban public security financial expenditures within a certain period of time	Negative
			The level of financial support for medical and health care	100 million yuan	Refers to urban medical and health financial expenditures within a certain period of time	Positive

(continued)

Table 2.2 (continued)

Target layer	Criterion layer	Sub-criterion layer	Indicator layer	Unit	Indicator description	Indicator type
			The level of financial support for transportation	100 million yuan	Refers to urban transportation financial expenditures within a certain period of time	Positive
		Material supply level	The comprehensive production capacity of water supply	Ten thousand cubic meters/day	Refers to the comprehensive production capacity calculated based on the design capabilities of water supply facilities such as water intake, purification, water delivery, and factory water main pipes	Positive
			The level of grain supply	Ten thousand tons	Refers to the total grain output of each city (state) within a certain period of time	Positive
			The level of emergency material supply	100 million yuan	Refers to the region's ability to quickly and effectively provide necessary supplies and services in emergencies or emergencies. It is specifically replaced by the total retail sales of consumer goods in each city (state)	Positive

Second, a random sampling of ten percent of these will be used to verify their conformity with the original data in Excel. Out of the 100 questionnaires that were distributed, nine of them were invalid. It means that 91% of the respondents actively participated in the survey.

(1) Reliability Test

A reliability test was conducted to evaluate the validity and reliability of the sample data, as well as the appropriateness of the relevant indices. Reliability refers to the degree of stability, consistency, and dependability of the results derived from the survey data. Stability refers to the degree of correlation between the outcomes achieved when experts conduct the survey multiple times, with intervals typically ranging from two to four weeks. The degree of agreement in the results is known as consistency. A high level of consistency shows a strong positive correlation between the responses provided by experts in the questionnaire. Reliability analysis involves several measures, such as internal consistency, test–retest reliability, and split-half reliability. This study predominantly employed internal consistency and reliability as it assessed the consistency of survey data and examined the correlation between survey questions and the measured content. Two commonly used approaches for assessing internal consistency are the split-half reliability coefficient and Cronbach's alpha coefficient. The half-reliability coefficient separates the questionnaire item into two pieces and calculates the scores for each portion separately. It then runs a correlation analysis on the scores to determine the correlation coefficient between the items. The coefficient refers to the reliability coefficient that measures the degree of consistency, specifically the half-reliability. The Cronbach's alpha coefficient is a quantitative measure that assesses the internal consistency of a psychometric test. It is commonly used as an indicator of its reliability. Lee Cronbach introduced Cronbach's alpha coefficient in 1951 as a statistical tool to evaluate and assess a scale's internal consistency and reliability that includes several items designed to measure a common underlying concept or construct. This study uses the Cronbach's alpha coefficient to carry out dependability testing. The function is as follows:

$$\alpha = \frac{K}{K-1}\left(1 - \frac{\sum_{i=1}^{K}\sigma_{Y_i}^2}{\sigma_X^2}\right) \tag{2.1}$$

where K represents the number of items on a scale, σ_X^2 is the total variance (i.e., the variance of the total scores given by respondents to the scale), and $\sigma_{Y_i}^2$ is the variance of the observed sample (i.e., the variance of the scores given by respondents to each item). Generally, a higher Cronbach's alpha coefficient indicates better reliability of the questionnaire. Cronbach's alpha values usually range from 0 to 1. Values below 0.6 indicate an unacceptable level of reliability, while values between 0.7 and 0.8 indicate an acceptable level of reliability. Values between 0.8 and 0.9 indicate a very good level of reliability (Nunnally 1978).

After importing the data into SPSS Statistics 27, the reliability test results are given in Table 2.3.

Table 2.3 Reliability test results

Criterion layer	Indicator layer	Cronbach's α	Cronbach's α after removing terms
Urban personnel safety resilience	C1	0.823	0.805
	C2		0.787
	C3		0.797
	C4		0.800
	C5		0.820
	C6		0.789
	C7		0.816
	C8		0.824
	C9		0.810
	C10		0.816
	C11		0.794
Urban facility safety resilience	C12	0.844	0.834
	C13		0.836
	C14		0.831
	C15		0.854
	C16		0.815
	C17		0.838
	C18		0.835
	C19		0.817
	C20		0.823
	C21		0.826
	C22		0.834
Urban management safety resilience	C23	0.752	0.689
	C24		0.746
	C25		0.735
	C26		0.717
	C27		0.741
	C28		0.692
	C29		0.719
	C30		0.736

The Cronbach's alpha values for urban personnel safety resilience, urban facility safety resilience, and urban management safety resilience are all greater than 0.7, as shown in Table 2.3. This means that the questionnaire is valid and meets the requirements for overall reliability. Also, the Cronbach's alpha values of each indicator after deleting items are less than the overall reliability. It means that the total reliability of the questionnaire decreases after deleting these indicators; thus, it is advisable to retain them.

(2) Validity Test

The validity was checked by conducting factor analysis on the internal structure of the data. This was referred to as the validity test. The PCA is mainly used to identify factors and evaluate the reliability of the structure. However, the results from exploratory factor analysis based on objective data rules may differ significantly from those from subjective experience. This validity test aimed to confirm the survey's validity rather than select an index. Structural equation modeling (SEM) was used to examine the objective selection of indices based on their level of interaction with one another.

A Kaiser–Meyer–Olkin (KMO) value less than 0.5 means the data is unacceptable for factor analysis. A KMO value ranging from 0.5 to 0.7 means the data is middling for factor analysis. A KMO value ranging from 0.7 to 0.9 means the data is meritorious for factor analysis. AKMO value greater than 0.9 means the data is marvelous for factor analysis (Deyin 2007). The significance level for Bartlett's Test of Sphericity should be less than 0.01. A KMO higher than 0.5 and significance less than 0.01, which means the results passed Bartlett's Test of Sphericity, indicates that the validity test was successful.

① Data Factor Analysis of the criterion "Urban Personnel Safety Resilience"

The importance evaluation data for each indicator under the "Urban Personnel Safety Resilience" criterion were organized and input into the SPSS software. The KMO and Bartlett's test results are shown in Table 2.4.

As can be seen in Table 2.4, the KMO value of urban personnel safety resilience is 0.872, and the significance is less than 0.001, indicating that the data is meritorious for factor analysis. After factor analysis of the data matrix, the corresponding rotated factor loading matrix is obtained, and the factor analysis table for the "urban personnel safety resilience" is shown in Table 2.5.

As can be seen in Table 2.5, there are three common factors: F1, F2, and F3. The cumulative variance explained variation value within these three common factors is 72.08%, which is greater than 50%, indicating that the information on the research item can be effectively extracted. The results show that these three common factors can well reflect the 11 variables under this criterion. The common factors

Table 2.4 Results of the criterion "Urban Personnel Safety Resilience" KMO and Bartlett's test

KMO and Bartlett test		
KMO sampling suitability quantity		0.783
Bartlett's test of sphericity	Approximate chi-square	600.727
	Degrees of freedom	55
	Significance	< 0.001

Table 2.5 "Urban Personnel Safety Resilience" evaluation factor analysis

Factor	Project description	Factor loading value matrix			Explained variation (%)
		1	2	3	
F1	The density of permanent residents in the built-up area	0.698	0.130	0.248	40.56
	The level of basic medical insurance for urban workers	0.881	0.118	0.296	
	The proportion of temporary residents	0.821	0.153	0.464	
	The number of newly added higher education talents	0.787	0.094	0.490	
F2	The level of reserve health technical personnel in the city	0.499	0.190	0.578	16.36
	The reserve level of urban health institutions	0.762	0.172	0.455	
	The level of social organization units	0.149	0.856	0.149	
F3	The number of people covered by unemployment insurance	0.281	0.807	0.179	15.16
	The number of people covered by urban old-age insurance	0.532	0.528	0.046	
	The number of people covered by work-related injury insurance	0.412	0.088	0.558	
	The per capita emergency deposit level	0.721	0.031	0.393	

F1, F2, and F3 correspond one-to-one with the sub-criterion level "basic population attributes," "social participation readiness," and "safety perception and safety culture," respectively, and can better represent the original 11 indicators.

② Data Factor Analysis of the criterion "Urban Facility Safety Resilience"

According to the data in Table 2.6, it can be concluded that the KMO value of urban facility safety resilience is 0.797, and the significance is less than 0.001, indicating that the data is meritorious for factor analysis. After factor analysis of the data matrix, the corresponding rotated factor loading matrix is obtained, and the factor analysis for the "urban facility safety resilience" is shown in Table 2.7.

As can be seen in Table 2.7, there are four common factors F1, F2, F3, and F4. The cumulative variance explained variation value within these four common

Table 2.6 Results of "Urban Facility Safety Resilience" KMO and Bartlett's test

KMO and Bartlett test		
KMO sampling suitability quantity		0.797
Bartlett's test of sphericity	Approximate chi-square	687.553
	Degrees of freedom	55
	Significance	< 0.001

Table 2.7 "Urban Facility Safety Resilience" evaluation factor analysis

Factor	Project description	Factor loading value matrix				Explained variation (%)
		1	2	3	4	
F1	Urbanization rate	0.521	0.515	0.301	0.462	44.16
	Proportion of land area in safety-vulnerable areas	0.520	0.274	0.182	0.683	
	The level of personnel reserves in the construction industry	0.518	0.621	0.307	0.076	
F2	Per capita urban road area	0.263	0.616	0.437	0.266	18.11
	The level of urban traffic lighting facilities	0.896	0.185	0.176	0.025	
F3	The penetration rate of wired broadcasting and television	0.486	0.214	0.689	0.168	9.38
	The level of urban drainage facilities	0.688	0.310	0.192	0.004	
F4	Per capita refuge area	0.904	0.311	0.080	0.079	7.36
	Building area of disaster relief reserve institution warehouses per ten thousand people	0.859	0.373	0.165	0.057	
	The number of beds in medical and health institutions per ten thousand people	0.816	0.372	0.216	0.081	
	The green coverage rate	0.494	0.579	0.008	0.090	

factors is 79.00%, greater than 50%, indicating that the information of the research item can be effectively extracted. The results show that these four common factors can well reflect the 11 variables under this criterion. The common factors F1, F2, F3, and F4 correspond to the sub-criterion "construction engineering," "transportation facilities," "lifeline engineering facilities," and "emergency support facilities," respectively, and can better represent the original 11 indicators.

③ Data Factor Analysis of the criterion "Urban Management Safety Resilience"

Table 2.8 "Urban Management Safety Resilience" KMO and Bartlett's test

KMO and Bartlett test		
KMO sampling suitability quantity		0.751
Bartlett's test of sphericity	Approximate chi-square	215.676
	Degrees of freedom	28
	Significance	< 0.001

As can be seen in Table 2.8, the KMO value of urban management security resilience is 0.751, and the significance is less than 0.001, indicating that the data is meritorious for factor analysis. After factor analysis of the data matrix, the corresponding rotated factor loading matrix is obtained, and the factor analysis for the "urban management safety resilience" is shown in Table 2.9.

As can be seen in Table 2.9, there are three common factors F1, F2, and F3. This significant cumulative variance indicates that a substantial proportion of the information within the study variables has been effectively extracted. The results demonstrate that these three factors aptly represent the eight variables under this criterion layer. Factors F1, F2, and F3 correlate directly with the sub-criterion "risk control level," "support and guarantee input level," and "material supply level," respectively, effectively embodying the original set of eight indicators.

Table 2.9 "Urban Management Safety Resilience" evaluation factor analysis

Factor	Project description	Factor loading value matrix			Explained variation (%)
		1	2	3	
F1	The death rate per million people due to disasters	0.798	0.360	0.007	38.43
	The proportion of annual direct economic losses due to disasters to the region's gross domestic product	0.268	0.846	0.021	
F2	The level of financial support for public safety	0.488	0.683	0.268	21.58
	The level of financial support for medical and health care	0.636	0.402	0.203	
	The level of financial support for transportation	0.566	0.330	0.678	
F3	The comprehensive production capacity of water supply	0.799	0.188	0.016	9.59
	The level of grain supply	0.652	0.000	0.305	
	The level of emergency material supply	0.581	0.329	0.317	

Reliability and validity tests conducted using SPSS software on the urban safety resilience coupling coordination index system have verified the system's high consistency, robust internal structure, and congruent hierarchical relationships among the indicators. Overall, the fit between the sub-criterion and indicator is high, exhibiting well-aligned structural relationships. This analysis reinforces the integrity of the index system and highlights its capacity to map and measure the dynamics of urban safety resilience precisely.

References

Deyin, F. 2007. Statistical testing issues in principal component analysis. *Statistics Education* 9: 4–7 (in Chinese).

Hongyan, S. 2019. Research of multiple planning integration index system. Hunan Normal University (in Chinese).

Hui, X. 2018. Construction of evaluation index system of campus sports environment in primary and middle S. Yangtze University (in Chinese).

Nunnally, J. C. 1978. *Psychometric Theory*. McGraw-Hill.

Ran, W. 2017. Study on establishment and demonstration of ecological civilization evaluation index system at the provincial level in China. China University of Geosciences (in Chinese).

Ruihua, Z. 2018. Empirical study on post evaluation index system of construction project based on AHP method. Xihua University. (in Chinese).

Wan, W., L. Ruiqi, X. Fengjiao, H. Hong, and Q. Tingxin. 2022. Interpretation of GB/T 40947-2021, guide for safety resilient city evaluation. *Standard Science* 2022 (11): 84–86+94 (in Chinese).

Weicheng, F. 2020. Trends in the development of safe and resilient cities. *Labour Protection* 3: 20–23 (in Chinese).

Ya, L., Z. Guofang, and G. Fumei. 2016. Review on methods of quantification of urban infrastructure resilience 23 (6): 113–122 (in Chinese).

Chapter 3
Measurement and Temporal Evolution of Urban Safety Resilience Coupling Coordination in the Chengdu-Chongqing Economic Circle

This chapter discusses the measurement and temporal evolution of urban safety resilience coupling coordination in the Chengdu-Chongqing Economic Circle. It begins with constructing a comprehensive model that integrates the previously developed urban safety resilience evaluation index system. The model comprises various components, including a weighting model that blends subjective and objective methods, a development level evaluation model, an obstacle degree model, and a coupling coordination model.

It details the entropy method for calculating weights, ensuring all indicators are standardized for comparability. Experts then determine the relative importance of these indicators. This combination of entropy and expert survey methods is used to finalize the weights for the comprehensive index system. Subsequently, the development level of urban safety resilience subsystems is evaluated using a linear weighting method, calculating comprehensive scores that indicate the developmental effectiveness of each subsystem. This evaluation forms the basis for analyzing system coupling and coordination. An obstacle degree model is introduced to identify the weaknesses and limiting factors in the construction of urban safety resilience, providing a statistical measure of obstacles to development. This model helps identify which factors hinder the achievement of optimal urban safety resilience, guiding targeted improvements.

The chapter culminates in constructing a coupling coordination degree model, which assesses the interaction and coordination between the subsystems of urban safety resilience. It categorizes the development types based on coupling coordination values, offering a nuanced view of the region's progress and challenges in enhancing urban safety resilience.

This analysis is vital for understanding the dynamics within the Chengdu-Chongqing Economic Circle, enabling policymakers and planners to address specific areas of concern effectively and enhance overall urban resilience.

Y. Xiang et al., *Research on Coupling Coordination of Urban Safety Resilience Based on Disaster Prevention and Reduction*, Resilient and Friendly Living Environments, https://doi.org/10.1007/978-981-96-1562-9_3

Methodology and Model Construction

Weight Calculation Methods

Weight calculation involves determining the importance level of indices. Subjective and objective weighting methods are commonly employed to determine weights. Subjective weighting methods rely on the decision-maker's emphasis on indices and are suitable for many hard-to-quantify indices. The most common subjective weighting methods are the analytic hierarchy process (AHP), comprehensive fuzzy evaluation, and expert survey method. However, subjective weighting methods may introduce significant bias. To address this problem, scholars have developed objective weighting methods, which consider the degree of variation in the concentration of attributes. They determine weights based on the amount of information in each attribute index. The common objective weighting methods include data envelopment analysis (DEA), entropy method, gray relational analysis (GRA), and principal component analysis (PCA).

(1) Entropy Method

According to Shannon's information theory, the entropy method can calculate weights by analyzing the information content of each index, minimizing the influence of subjective judgments. It is an objective weighting method commonly used in multi-index comprehensive evaluation.

(2) Analytic Hierarchy Process

AHP is a method for quantifying qualitative problems. It turns qualitative evaluations into data through a hierarchical structure. It is suited for complex decision-making.

(3) Expert Survey Method

Expert survey method involves consulting experts or authoritative figures about the importance of indices, usually through independently filled scales to select weights, which are then statistically analyzed to determine the weights of indices.

(4) Gray Relational Analysis

The GRA was proposed by Professor Deng Julong in 1985, and it is used to analyze the degree of correlation between factors. It constructs a gray system comprising the phenomena and their influencing factors. It uses the degree of similarity in the geometric shapes of sequence curves to determine the correlation between factors. The level of correlation directly affects the relationship between the indicators and the comprehensive evaluation object. A higher correlation indicates a more significant impact of the indicator on the comprehensive evaluation object, resulting in a higher weight value.

(5) Principal Component Analysis

The statistician British introduced the PCA, which recombines several correlated indices into a smaller number of composite indices that reflect the original variables' information. They calculate the coefficients of indices in each principal component's linear combination and their variance contribution rates to normalize the weights of indices.

(6) Data Envelopment Analysis

The DEA method was proposed by American operations researchers Charnes and Cooper. DEA evaluates relative effectiveness based on input and output indices without changing the inputs or outputs of decision units, using statistical data to compare the deviations of decision units.

The advantages and disadvantages of the weight calculation methods are summarized in Table 3.1.

As seen in Table 3.1, subjective and objective methods have advantages and disadvantages due to different emphases. Although the subjective methods are flexible, they rely too much on the evaluator's subjective judgment and lack a scientific quantitative basis. In contrast, the objective methods can reduce the impact of personal bias, and they may be affected by the fluctuation range of the value itself, the degree of variation, and the contribution of the data. The true importance of the indicator is not always fully reflected.

Models of Coupling Effects

Society constitutes a multifaceted system encompassing various interdependent domains, including resource management, environmental conservation, and economic activities. With a better grasp of scientific development, evaluating a region's or a society's progress has transitioned from focusing exclusively on absolute economic growth metrics to a broader assessment of developmental harmony and balance. The concepts of coupling effects and the degree of coordinated development have become essential metrics for gauging these dimensions.

In exploring coupling effects, a diverse array of models is employed, such as the Environmental Kuznets Curve (Dasgupta et al. 2002), bi-exponential models (Jinchuan and Chuanglin 2003), nonlinear dynamics models (Jie et al. 2010), coupling degree models (Xisheng et al. 2013), gray relational analysis (Xuemei et al. 2014), and dynamic coupling models (Biao and Chuanglin 2012). Each model is constructed for specific scenarios, with distinct advantages and limitations. The coupling degree model elucidates the dynamic interactions among different system components, while the dynamic coupling model concentrates on evolving interaction patterns. In addition, advanced statistical tools like vector autoregression and spatial regression models are instrumental in analyzing temporal and spatial dependencies within data.

Table 3.1 The advantages and disadvantages of the weight calculation methods

Category	Method	Advantage	Disadvantage
Subjective methods	AHP; Expert survey method	① Systematic analysis method, especially suitable for systematic evaluation without structural characteristics; ② Simple and practical, transforming multi-objective and difficult-to-quantify decision-making problems into multi-level single-objective problems; ③ Less quantitative data required	① When there are too many indicators, the workload is heavy, and the weight is difficult to determine; ② In the actual operation process, there is excellent subjectivity and arbitrariness, with a strong dependence on the identification and judgment of the decision-maker. When the decision-maker disagrees on the same issue, if the values are too large, the accuracy of the evaluation will be affected
Objective methods	DEA	There is no need to consider the identity of dimensions with strong objectivity	Determine the weight from the perspective of the most favorable evaluation unit. The outstanding performance of a specific data may distort the evaluation results, causing the weight calculation results to be inconsistent with the actual situation
	Entropy method	The method has been widely used to determine the weight according to the degree of variation of the indicator. It can avoid errors caused by human factors	① Unable to reduce the index dimension; ② High dependence on data
	Gray relational analysis method	There is no excessive requirement on the number of samples. It is suitable for gray systems that deal with incomplete data and asymmetric information	① The theoretical system of the GRA is not perfect. Due to the small amount of data, it is difficult to analyze the relationship between factors clearly; ② The gray system limits the GRA, and the optimal value of the indicator needs to be determined during the solution process., there is a certain degree of subjectivity
	Principal component analysis	Dimensionality reduction calculation of indicators reduces information loss and can intuitively reflect the indicators that have a greater impact on the evaluation results, with strong objectivity	① The integrity of the original data is required to be high; ② Obtaining the weight of the limited principal components cannot directly obtain the weight of each independent indicator

These research models have unique strengths and limitations; their specific applicability is shown in Table 3.2.

Researchers can utilize these methods to quantitatively analyze and evaluate the degree of synergistic development within and between systems. It provides essential methodologies for societal and industrial advancement. These measurement standards clarify the interactions within a system and provide guidance for enhancing and optimizing the inherent coupling of multiple systems.

Coupling Coordination Measurement Models

The coupling degree model can be well recognized and highly regarded among the several models used to measure coupling effects. Its popularity stems from its straightforward computing process and easily understandable outcomes. It serves as a tool to evaluate interaction levels across diverse domains such as the environment (Wei and Yongfeng 2012; Honghao and Xiaohong 2019; Yang et al. 2020a), economy (Gang et al. 2017), urbanization (Xiaomin et al. 2022; Ariken et al. 2020; Ma et al. 2022), agriculture (Yang et al. 2020b), industry (Yan and Yuming 2006), transportation (Deyou et al. 2013), and demographics (Fawei et al. 2014; Jilai et al. 2018). The coupling coordination degree model, which is an enhanced version of this model, is commonly used to measure the complexity of inter-system coupling effects. The coupling coordination degree model leverages the coupling degree to delineate interrelations between subsystems and incorporates the degree of coordinated development to evaluate the entire system comprehensively. Both the coupling degree and coordination degree metrics quantify the level of coordination between systems, thereby elucidating the extent of their synergistic development (Chuansong et al. 2018; Weibing and Tanling 2017).

The coupling degree quantifies the intensity of interactions within and between subsystems or elements of a system, thereby reflecting the synergistic transition from disorder to order within the system. In contrast, the coordination degree assesses the harmony among elements within and between systems. This phenomenon reflects the positive cyclic interaction between the system and its coupling elements and the transition from disorder to progress, laying the foundation for the stable and healthy development of multiple systems and their elements. Essentially, the coupling degree delineates the strength of interactions among multiple objects but does not assess the benign characteristics of connections between systems. On the other hand, the coordination degree, which is rooted in the coupling degree, quantifies the quality of coordination conditions. This book aims to comprehensively analyze the many coupling coordination models proposed by the academic community to determine more appropriate measures for coupling coordination.

(1) Ratio Method

The ratio analysis constructs a model for system coupling coordination primarily through proportional forms. Cao Lijun et al. utilized this method to develop a model

Table 3.2 Summary of coupling effects models

Model	Advantages	Disadvantages	Applicable scope
Environmental Kuznets Curve	A succinct exposition of the correlation between economic expansion and the utilization of resources	Simplified complex economic resource relationships, unable to fully reflect the actual situation	Appropriate for acquiring a basic comprehension of the overarching correlation between economic expansion and the use of resources
Double exponential model	Flexibly applicable to describe specific growth or decline trends	In some cases, it may be overly simplified and might not fully capture the changing trends	Suitable for preliminary trend forecasting, but may not be precise enough for complex changes
Nonlinear dynamical model	Capturing nonlinear changes and complex dynamic relationships within the system	The modeling process is complex, making it difficult to explain and validate the results	Suitable for simulating nonlinear dynamic changes within complex systems
Coupling degree model	Assessing the degree of interconnection among internal components of the system, revealing the interactions within complex systems	Demands high data quality, posing challenges in interpreting model results	Suitable for analyzing the interconnectivity among various components within complex systems
Dynamic coupling model	Capturing the temporal evolution of the system and interactions among its components	Complex modeling requires high-quality data and parameters, making interpretation and validation challenging	Suitable for simulating the mutual interactions between various components in dynamic systems
Vector autoregressive model	Applicable for time-series data modeling, considering interactions among variables	Requires high data stationarity and linearity for the analysis	Used for analyzing dynamic relationships between variables in time-series data

(continued)

Table 3.2 (continued)

Model	Advantages	Disadvantages	Applicable scope
Spatial regression model	Considering spatial correlation, applicable for describing spatial data relationships	Requires a high level of spatial autocorrelation, posing challenges in interpretation and result validation	Suitable for analyzing spatial data and its spatial correlations
ArcGIS centroid curve optimization grading	Optimizing and grading geographic spatial data in a GIS environment	Parameter settings and data precision influence results; cautious use and interpretation are necessary	Used for geographic spatial data analysis and visualization, facilitating the understanding of geographical distribution patterns

for regional sustainable development's coordination, quantitatively evaluating the coordination degree of regional population, resources, environment, and economic subsystems (Lijun et al. 2000). The formula is as follows:

$$D = K\frac{q \cdot \frac{E}{E_0}}{(1-q) \cdot \left(\frac{N}{N_0}\right)\left(\frac{R}{R_0}\right) \cdot e^{\frac{P}{P_0}}} \tag{3.1}$$

where D represents the coordination development degree; K denotes the adjustment coefficient; q signifies the Engel coefficient for the evaluation year; E and E_0 represent the economic indices for the evaluation year and the base year; N and N_0 denote the population indices for the evaluation year and the base year; R and R_0 signify the resource indices for the evaluation year and the base year; P and P_0 represent the environmental indices for the evaluation year and the base year.

The ratio evaluation method provides a clear structure and simplicity in the calculation. However, its dependence on the experiential judgment of researchers may pose challenges in accurately assessing coupling coordination under various scenarios, imposing significant limitations on its application scope. Therefore, this method is less frequently employed.

(2) Elasticity Coefficient Method

In the field of economics, the elasticity coefficient represents the ratio of the change rate of an economic indicator x to the change rate of the economic indicator y during a specific period. It illustrates the dependent relationship between the extent of changes in one variable and the extent of changes in another related variable. The Elasticity Coefficient Method primarily utilizes this concept to formulate a model for assessing the coupling degree between systems. Yansui and Yurui (2010) employed the Elasticity Coefficient Method to develop a coupled model for changes in arable land and agricultural labor. They revealed the spatiotemporal coupling characteristics between the area of arable land and variations in agricultural labor, with their labor-cultivation elasticity coefficient model expressed as follows:

$$E = \frac{\text{LR}}{\text{FR}} = \frac{(L - L_0)/L_0}{(F - F_0)/F_0} \tag{3.2}$$

where LR and FR represent the rates of change in agricultural labor and arable land area; L and L_0 denote the quantities of agricultural labor for the evaluation year and the base year; F and F_0 signify the arable land areas for the evaluation year and the base year; E denotes the labor-cultivation elasticity coefficient, reflecting the degree of mutual influence between changes in agricultural labor and arable land area.

According to the labor-land elasticity coefficient, Ma Li and colleagues devised an elasticity coefficient evaluation model to analyze changes in agricultural labor and agricultural economic development (Li et al. 2018). Cui Panpan et al. established an elasticity coefficient model to assess the ecological environment of the Yellow River Basin and the level of high-quality development (Panpan et al. 2020). Based on the elasticity coefficient method, these models represent a further advancement and expansion of the ratio evaluation method. These findings reveal the spatiotemporal characteristics of system coupling and provide insights into the direction and relative pace of system changes. Although this method effectively quantifies the coupling relationships between systems, it does not adequately assess their coordination degrees.

(3) Efficacy Function Method

The efficacy function method is based on the synergy theory and the efficacy function. Its main purpose is to develop coupling coordination models. The calculation procedure is quite simple. Liu Yaobin analyzed the link and harmony between urbanization and the environment in China (Yaobin et al. 2005). Ma Xiaoqin measured the interdependence between rural tourism and rural revitalization (Xiaoqin 2019). Their model integrates both an efficacy function and a coupling coordination function. Assuming $U_i(i = 1, 2, \ldots, n)$ represent the ordinal parameter of the coupled comprehensive system, and $X_i(i = 1, 2, \ldots, n)$ denote the value of the i-th ordinal parameter. $\text{Max}(X_i)$ and $\text{Min}(X_i)$ correspond to the high and low extreme values of the ordinal parameters at stable critical points for the coupled system. The ordered effectiveness function for each subsystem in the coupled system is given by:

$$U_i = \begin{cases} \frac{X_i - \text{Min}(X_i)}{\text{Max}(X_i) - \text{Min}(X_i)} & \text{(positive effect)} \\ \frac{\text{Max}(X_i) - X_i}{\text{Max}(X_i) - \text{Min}(X_i)} & \text{(negative effect)} \end{cases} \tag{3.3}$$

where U_i represents the effectiveness function, showing how satisfied the indicator is with the target value. The range of U_i is [0,1], where the value closer to 1 indicates higher satisfaction and the value closer to 0 indicates lower satisfaction. The coordination function can be established by utilizing the geometric mean and linear weighting. These methods determine the coupling coordination degree within the system.

The function of the coupling coordination function using the geometric mean method is as follows:

$$C = \sqrt{U_1 \times U_2 \times \cdots U_n} = \sqrt{\prod_{i=1}^{n} U_i} \tag{3.4}$$

The ordinal parameter of each subsystem has a distinct impact on the system. It is necessary to allocate appropriate weights. The coupling function using the linear weighting method is constructed as follows:

$$C = w_1 \cdot U_1 + w_2 \cdot U_2 + \cdots w_n \cdot U_n = \sum_{i=1}^{n} w_i \cdot U_i, \sum_{i=1}^{n} w_i = 1 \tag{3.5}$$

where U_i represents the effectiveness or comprehensive ordinal parameter of the subsystem; w_i represents the weight of each ordinal parameter in the system; C represents the coupling coordination of the system with the range [0,1]. The higher value of C represents more coupling among subsystems, meaning the stronger interactions and a tendency for the system to develop in an organized direction; the lower value C represents less coupling among subsystems, meaning the looser, disordered relationships and development.

(4) Spatially Discrepancy Coefficient Method

The spatially varying coefficient models are based on minimizing geometry discrepancies. It compares the differences among multiple systems to make a model for coupling coordination evaluation (Xiaonan 2019). One notable model is the Urban-Travel Coupling Model proposed by Xiangyang et al. (2021). The model is as follows:

$$C = \frac{\text{TE} \times \text{HST}}{[(\text{TE} + \text{HST})/2]^k} \tag{3.6}$$

where TE represents urban tourism efficiency; HST represents the advantage degree of high-speed transportation; k is the adjustment coefficient. In this case, $k = 2$ due to the two subsystems; C is the connection between tourism efficiency and the advantage of high-speed transport. Another model is the coupling degree model. It is for the three main systems of regional economy, ecological environment, and tourism. Cheng et al. (2016) established it. The formula for this model is as follows:

$$C = \left\{ \frac{J \times H \times L}{[(J + H + L)/3]^3} \right\}^{1/3} \tag{3.7}$$

where J, H, and L represents the three subsystems of economic development, ecological environment, and tourism industry; C represents the coupling degree of the three subsystems with the range from 0 to 1. Another model established by Zhao Lei et al. is the coupling degree function for the tourism industry and new urbanization system, which is shown as follows (Lei et al. 2020):

$$C = 2 \times \sqrt{(U_1 \cdot U_2)/\prod(U_1 + U_2)} \tag{3.8}$$

where C represents the system coupling degree with a range from 0 to 1, which depends on the ordinal parameters of the tourism industry and new urbanization. Coupling is a cross-sectional concept that measures the coupling degree between the tourism industry and new urbanization in a specific period. However, in some situations, the coupling degree may not show the overall effectiveness, and synergy between the tourism industry and new urbanization. Therefore, it is necessary to add a coordination function to the system on top of coupling. It will measure how well the tourism industry and the new urbanization subsystem work together in different evolutionary stages (Lei et al. 2020). The function of the tourism industry-new urbanization coupling coordination function is as follows:

$$D = \sqrt{C \times T}, T = \alpha \times U_1 + \beta \times U_2 \tag{3.9}$$

where D represents the coupling coordination degree between the tourism industry and new urbanization; T represents the comprehensive harmony index between the tourism industry and new urbanization, reflecting their synergistic effects; α and β are undetermined coefficients with $\alpha + \beta = 1$. It reflects the importance of each subsystem in the coupled comprehensive system. Specific values can be determined based on practical experience or expert opinions. Mei et al. (2021) introduced the relative development coefficient E, which can measure and analyze the relative development degree of the coupling coordination. This coordination is between the transformation of economic growth momentum and green development in 31 Chinese provinces and municipalities. They further enriched the evaluation model for coupling coordination, which is shown as follows:

$$C = \sqrt{\frac{f(x) \times g(x)}{\left[\frac{f(x)+g(x)}{2}\right]^2}} \tag{3.10}$$

$$D = \sqrt{C \times T}, T = \alpha f(x) + \beta g(x) \tag{3.11}$$

$$E = \frac{f(x)}{g(x)} \tag{3.12}$$

where $f(x)$ and $g(x)$ are the indices for economic momentum and green development; C, T, and D are the coupling, development, and coordination degrees; E is the relative development degree for economic momentum and green development. The judgment criteria are: if $0 < E \leq 0.8$, signifies a delay in momentum transformation, which restricts green development; if $0.8 < E < 1.2$, it indicates synchronous development; if $E \geq 1.2$, it suggests that green development lags, restricting momentum transformation.

The analysis above shows that the principle of constructing the coupling coordination model based on the geometric discrepancy coefficient is relatively simple.

The system incorporates four measuring indicators: the coupling degree C, the development degree T, the overall coordination degree D, and the relative development degree E. These parameters have been widely applied in various fields, including natural sciences and social sciences, regional sustainable development, urbanization construction, ecological environment, and tourism industry development.

(5) Gray Relational Method

The gray relational analysis (GRA) with coupling coordination degree is grounded in the theory of gray systems. It investigates the degree of similarity or dissimilarity in the development of internal elements of a system to explore their relevance and significant influencing factors. The analysis helps examine the system's coupling state and development trends. The main steps of the algorithm are as follows:

The GRA needs two sets of analysis sequences: First, assume two subsystems are in the coupled system. The sets are the subsystem sequence set (X_i) and the subsystem sequence set (Y_i). Standardize the data of various observed indicators within the system. Second, the correlation coefficient and relational degree between the subsystems are calculated. The correlation coefficient is as follows:

$$\xi_{ij}(t) = \frac{\min\limits_{t}\min\limits_{t}\left|Z_i^x(t) - Z_j^y(t)\right| + \rho \max\limits_{i}\max\limits_{j}\left|Z_i^x(t) - Z_j^y(t)\right|}{\left|Z_i^x(t) - Z_j^y(t)\right| + \rho \max\limits_{i}\overset{y}{\max\limits_{j}}\max\limits_{j}\left|Z_i^x(t) - Z_j^y(t)\right|} \tag{3.13}$$

where $\xi_{ij}(t)$ represents the correlation coefficient between subsystems i and j at time t, indicating the standardized values of various indicators for subsystems i and j at time t; ρ is the resolution, which enhances the significance of differences in correlation coefficients. It reflects the influence of each indicator on the degree of correlation and takes values between 0 and 1. Once the correlation coefficients for the system are obtained, the average is taken over the sample size k to construct an $m \times n$ relational matrix Ψ, reflecting the spatiotemporal interleaving relationships between subsystems. The relational matrix Ψ is as follows:

$$\psi = \begin{vmatrix} \psi_{11} & \cdots & \psi_{1n} \\ \cdots & \psi_{ij} & \cdots \\ \psi_{mn} & \cdots & \psi_{mn} \end{vmatrix}, \quad \psi_{ij} = \frac{1}{k}\sum_{i=1}^{k}\xi_{ij}(t) \tag{3.14}$$

where m and n are the numbers of indicators for subsystems i and j; k is the number of research samples; Ψ_{ij} is the element of the relational matrix. The closer it is to 1, the stronger the correlation between the systems; conversely, the closer Ψ_{ij} is to 0, the weaker the correlation between the systems.

Based on the correlation coefficients and the relational matrix, a gray relational coupling degree model can be constructed to assess the overall coupling status between systems. The model is as follows:

$$C(t) = \frac{1}{m \times n} \sum_{i=1}^{m} \sum_{j=1}^{n} \xi_{ij}(t) \tag{3.15}$$

where $C(t)$ represents the coupling degree of the system, with a range of [0, 1]. If $C(t) = 0$, the coupling degree is minimal, indicating that the subsystems are unrelated. If $C(t) = 1$, the coupling degree is maximal, indicating the most robust interaction between subsystems.

The model c takes into account both temporal and spatial dimensions. The approach has certain advantages in dealing with uncertainties such as "small sample size" and "limited information." However, it may not effectively address the information overlap caused by the correlation between evaluation indicators.

(6) Modified Models

According to the research proposed by Shujia et al. (2021), the model has been modified. The model has superior validity compared to traditional coupling coordination models. But it still needs to solve the subjectivity in index construction, volatility of coupling results, and unreliability in comparability. The model is shown as follows:

$$C = \sqrt{\left[1 - \frac{\sum_{i>j,j=1}^{n} \sqrt{\left(U_i - U_j\right)^2}}{\sum_{m=1}^{n-1} m}\right] \times \left(\prod_{i=1}^{n} \frac{U_i}{\max U_i}\right)^{\frac{1}{n-1}}} \tag{3.16}$$

$$T = \sum_{i=1}^{n} \alpha_i \times U_i, \sum_{i=1}^{n} \alpha_i = 1 \tag{3.17}$$

$$D = \sqrt{C \times T} \tag{3.18}$$

where $U_i \in [0, 1]$, $C \in [0, 1]$. If each subsystem is more discrete, the value of C is lower. The model makes C values more distinct, resulting in more efficient evaluation in social science. The calculated degree of coordinated development can more accurately represent the measurement of coupling coordination and development level.

Construction of the Urban Safety Resilience Coupling Coordination Model

1. Construction of the Weighting Model

This study combines subjective and objective approaches to determine the index weights. The entropy approach is commonly employed to analyze the weight allocation in complicated, nonlinear dissipative structural systems (e.g., urban safety resilience), which are influenced by several factors that determine their growth levels (Yunfei et al. 2021). Nevertheless, data may significantly influence objectively determined weights, necessitating careful consideration of various factors. Once the specific index weights have been determined, it is imperative to utilize the AHP and seek input from experts to establish the final weights for each criterion layer and make any necessary refinements to the specific index weights.

(1) Entropy Weight Method

① Normalize the Decision Matrix

Utilizing the optimal range standardization method, both positive and negative indicators of urban safety resilience evaluation are standardized to eliminate dimensional impacts and ensure data comparability.

$$\text{Positive Indicator: } x'_{\lambda ij} = \left(x_{\lambda ij} - x_{\min}\right)/(x_{\max} - x_{\min}) + A \tag{3.19}$$

$$\text{Negative Indicator: } x'_{\lambda ij} = \left(x_{\max} - x_{\lambda ij}\right)/(x_{\max} - x_{\min}) + A \tag{3.20}$$

where $x'_{\lambda ij}$ represents the dimensionless standardized value of the indicator after standardization; $x_{\max}$ and $x_{\min}$ are the maximum and minimum values of the original indicator values $x'_{\lambda ij}$; A is a small positive constant, typically set to 0.0001, added to each element of the denominator to prevent it from being zero.

② Calculating the Entropy for Each Indicator

$$E_j = -k\sum_{\lambda=1}^{t}\sum_{i=1}^{m} P_{\lambda ij}\ln P_{\lambda ij} \tag{3.21}$$

where k is a constant given by $k = 1/\ln(t \times m)$, $P_{\lambda ij}$ represents the weight of the j-th indicator in the i-th region. E_j indicates the information utility value of the indicator.

③ Calculating the Weights of Each Indicator

$$W_j = \left(1 - E_j\right)/\sum_{j=1}^{n}\left(1 - E_j\right) \tag{3.22}$$

(2) Expert Survey Method to Obtain the Relative Importance Value of Indicators

The relative importance values of indicators will be obtained through the expert survey method. The method can assess the relative importance of indicators. It involves inviting a certain number of experts to compare the relative importance of indicators within each level and derive numerical values for their relative importance. Since there are no extreme cases of extreme importance or unimportance among the indicators, a scale ranging from one to five will be used to gauge their relative importance. Experts will use their own experience to assess the relative importance of indicators and quantify the results, which will then be entered into the corresponding table. Refer to Appendix C for the questionnaire regarding the relative importance of evaluation indicators for urban safety resilience coupling coordination. The survey will be conducted using a combination of online methods (e.g., email and offline) and offline methods (e.g., interviews).

To ensure the effectiveness of the weighting results, the selection of surveyed experts will primarily consider the following aspects:

① Expert Qualifications

The selected experts should be professionals engaged in urban safety, urban infrastructure construction and management, disaster prevention and mitigation management, and urban emergency management.

② Determination of Experts' Number

In order to guarantee the reliability and functionality of the data collected from expert surveys, this study will invite 15 experts to assess the indicators and create the judgment matrix.

③ Determination of Expert Sources

When inviting experts to rate the relative importance, caution will be exercised to exclude experts who have already taken part in surveys, reducing the subjectivity of the results. Experts will be selected from different perspectives on evaluating urban safety resilience to obtain objective and scientific evaluation results. Considering the various stakeholders and beneficiaries involved in evaluating urban safety resilience coupling coordination, experts will primarily be selected from emergency management departments, universities and research institutions, and some enterprises and institutions.

(3) Calculation of Indicator Weight Based on AHP

A comprehensive survey was undertaken among specialists from varied disciplines in order to determine the relative significance of various indices. The study utilizes the square root method within the AHP framework. It combines decomposition with synthesis, interweaving subjective judgments with a rigorous quantitative and qualitative analysis to facilitate quantified decision-making.

The essential procedures in deploying AHP to determine the weights of evaluation indicators for subsystems in urban safety resilience are as follows:

① Construction of a Hierarchical Structure

A thorough examination of the evaluation object allows for its dissection into multiple levels, making a well-organized hierarchical structure. The hierarchical structure represents previously established evaluation indicators for various subsystems of urban safety resilience. The decision-making objective corresponds to the criterion layer, which consists of the subsystems of urban safety resilience. The intermediary elements constitute the sub-criterion layer, and alternative options form the indicator layer.

② Construction of Judgment Matrix

The judgment matrix can be constructed through systematic scoring by experts, who assess the internal importance of each index within the tiers.

③ Calculation of Final Weights

This involves computing the eigenvalues and eigenvectors of the judgment matrix. The eigenvectors indicate the ordered weights of the indices. The judgment matrix's consistency is carefully examined to ensure it meets the required standards, primarily by a random consistency test of the eigenvalues. Generally, the consistency ratio should be less than 0.1. The judgment matrix must be recalculated if the threshold is exceeded until the consistency ratio is satisfied.

The calculation process of the AHP is conducted through Yaahp 10.1 software. The weights derived from the entropy method and those obtained from the AHP are averaged to produce a comprehensive weight for the indicator system, which assesses the urban safety resilience coupling coordination.

2. Development Level Evaluation Model

(1) Subsystem Development Level Evaluation Modeling

The evaluation model for the subsystem development level can compute the overall scores of each subsystem, measuring their development effectiveness. This model serves as the foundation for determining system coupling and coordination. The common evaluation methods include the single-indicator model and the multi-indicator model.

The single-indicator model involves comparing individual indicators and is suitable for numerical analysis when the number of indicators is small. The multi-indicator model incorporates all indicators into the overall system. By analyzing the combined impact of multiple indicators on the evaluated object. It calculates and analyzes the development status of the system as a whole.

This study selects the linear weighting method from the comprehensive index evaluation method to calculate the development level evaluation values. It has been made to ensure the evaluation structure's comprehensiveness. The evaluation model for the criterion layer of urban safety resilience is given in Eq. 3.23.

$$U_m = \sum_{j=1}^{n} w_j x'_{\lambda ij} \tag{3.23}$$

where $m = 1, 2, 3$; U_1, U_2, and U_3 represent the values of urban personnel safety resilience, urban facility safety resilience, and urban management safety resilience; $U_1 \in [0, 1]$, $U_2 \in [0, 1]$, and $U_3 \in [0, 1]$; $x'_{\lambda ij}$ denotes the standardized value of the j-th indicator under each of the three subsystems, while w_j represents the corresponding weight of the indicator.

It is evident that the values of U_1, U_2, and U_3 are directly proportional to the developmental level of the subsystems. As the values of U_1, U_2, and U_3 increase, the corresponding levels of urban personnel safety resilience, urban facility safety resilience, and urban management safety resilience increase; and vice versa.

(2) Comprehensive System Development Level Modeling

Building upon the comprehensive evaluation values of subsystems' developmental levels, the evaluation value of the comprehensive system developmental level can be estimated. It is used to assess the overall development status of the urban safety resilience comprehensive system, which is given in Eq. 3.24.

$$T = \alpha_1 U_1 + \alpha_2 U_2 + \alpha_3 U_3 \tag{3.24}$$

where T represents the evaluation value of the comprehensive system's developmental level within the interval $T \in [0, 1]$. A higher value of T indicates a higher developmental level of the comprehensive system. The coefficients α_1, α_2, and α_3 reflect the undetermined coefficients of for urban personnel safety resilience, urban facility safety resilience, and urban management safety resilience systems, respectively. They reflect the extent to which subsystems influence the subsystems' impact on the comprehensive system is reflected by the constraint $\alpha_1 + \alpha_2 + \alpha_3 = 1$. While the specific values of α_1, α_2, and α_3 may have some impact on the outcomes, their impact on the overall development trend is relatively insignificant. The influences of urban personnel safety resilience, urban facility safety resilience, and urban management safety resilience on the comprehensive system's development are equally important. Based on the coupling mechanism between the two systems, it is known that there is a degree of mutual compensation between systems. Hence, this study sets $\alpha_1 = \alpha_2 = \alpha_3 = 1/3$, which leads to the transformation of the equation as follows:

$$T = \frac{U_1 + U_2 + U_3}{3} \tag{3.25}$$

3. Obstacle Degree Model

The obstacle degree model is introduced to identify the shortcomings and constraints in constructing urban safety resilience. The model is a mathematical model used to statistically quantify the obstacles that impede the progress of various entities. It identifies the degree of obstacles to individual indicators by calculating the contribution and deviation of each indicator factor in the evaluation indicator system. It allows recognition of how much each unidirectional indicator hinders the overall evaluation goal (Zhenjie et al. 2020). It provides theoretical guidance for adopting targeted measures according to local conditions. The specific calculation steps are as follows:

(1) Determine Factor Contribution

Factor contribution refers to the impact level of a single indicator on each subsystem of urban safety resilience, represented by the weight W_j.

(2) Calculate Indicator Deviation

Indicator deviation is the distance between each indicator and its optimal value. It is represented by the difference between the standardized value of the individual indicator:

$$d_{\lambda ij} = 1 - x'_{\lambda ij} \tag{3.26}$$

where $d_{\lambda ij}$ represents the indicator deviation of the j-th indicator in the i-th region in the λ-th year, and $x'_{\lambda ij}$ represents the standardized value of each indicator.

(3) Calculate $O_{\lambda ij}$

Calculate the hindrance degree $O_{\lambda ij}$ of each individual indicator to the development level of each subsystem of urban safety resilience using the following formula:

$$O_{\lambda ij} = \frac{d_{\lambda ij} W_j}{\sum_{j=1}^{n} d_{\lambda ij} W_j}, \quad S_{\lambda i} = \sum_{j=1}^{n} O_{\lambda ij} \tag{3.27}$$

where $O_{\lambda ij}$ represents the hindrance degree of the j-th indicator in the i-th region in the λ-th year. A higher value indicates a stronger constraint of this indicator on the development of various subsystems of urban safety resilience. $S_{\lambda i}$ represents the hindrance degree at the subsystem level.

4. Coupling Coordination Degree Modeling and Development Type Classification

(1) Coupling Coordination Degree Modeling

Based on the modified model proposed by Wang Shujia et al., the coupling coordination degree model is as follows:

Where max U_i is the maximum value of U_3,

$$C = \sqrt{\left[1 - \frac{\sqrt{(U_3 - U_1)^2} + \sqrt{(U_2 - U_1)^2} + \sqrt{(U_3 - U_2)^2}}{3}\right] \times \sqrt{\frac{U_1}{U_3} \times \frac{U_2}{U_3}}} \tag{3.28}$$

$$D = \sqrt{C \times T} \tag{3.29}$$

The main benefit of this model is its capacity to distribute C evenly as much as possible across the range from 0 to 1. It can increase the distinguishability of C values, resulting in improved efficiency in social science. Based on the modified model, the subsequently calculated degree of coordinated development can more reasonably represent the coupling coordination and development level measurement.

(2) Classification of Coupling Coordination Development

Existing research employs diverse methodologies to classify coupling coordination types, primarily through percentile division based on coupling coordination values or direct subjective division. The percentile method is strictly divided based on the statistical information objectively provided by the data. However, this method lacks consistency in the criteria for classifying various datasets, making it challenging to compare different studies. Following the objective criteria to establish a standardized system for identifying different coupling coordination development types is necessary. It can be done by artificially grouping them based on the development of coupling coordination degree. Moreover, a more precise classification can be achieved by contrasting disparities in assessment results for each subsystem.

Most current classification methods are typically designed for coupling coordination models involving only two systems. When dealing with models with three or more systems, classifications typically only determine major categories without additional subdivisions. Considering the theme and significance of the research, as well as referencing previous research achievements and making improvements, a classification of stage types for urban safety resilience coupling coordination is proposed, as shown in Table 3.3.

Table 3.3 classifies urban safety resilience coupling coordination development stages in different regions. The classification is based on the evaluation scores of urban safety resilience coupling coordination and its three subsystems: urban personnel safety resilience, urban facility safety resilience, and urban management safety resilience. The classification includes eight major categories, amounting to 32 specific types. It is essential to highlight that if a specific type of coupling coordination development in a particular region satisfies two out of the three lagging types (i.e., urban personnel safety resilience, urban facility safety resilience, and urban management safety resilience), a further comparison of U_1, U_2, and U_3 is required. If the difference between the two smaller values among U_1, U_2, and U_3 is less than 0.1, the region is considered to belong to both of these two lagging types; otherwise, it belongs to the type that simultaneously satisfies the two conditions in the specific type classification criteria. For instance, if the scores for region A are $U_1 = 0.8$, $U_2 = 0.4$, $U_3 = 0.35$, and $U_2 - U_3 < 0.1$, region A is classified as the urban facility safety resilience lagging type and the urban management safety resilience lagging type. Conversely, if the scores for region B are $U_1 = 0.8$, $U_2 = 0.4$, $U_3 = 0.25$, and $U_1 - U_2 \geq 0.1$, $U_1 - U_3 \geq 0.1$, $U_2 - U_3 \geq 0.1$, region B is classified as the urban management safety resilience lagging type.

Table 3.3 Division of development stages of coupling coordination degree

Category	Division basis	Concrete type	Specific type classification basis
High-quality coordinated development	$0.8 \leq D < 1$	High-quality coordinated development with lagging urban personnel safety resilience	$U_2 - U_1 \geq 0.1$ or $U_3 - U_1 \geq 0.1$
		High-quality coordinated development with lagging urban facility safety resilience	$U_1 - U_2 \geq 0.1$ or $U_3 - U_2 \geq 0.1$
		High-quality coordinated development with lagging urban management safety resilience	$U_1 - U_3 \geq 0.1$ or $U_2 - U_3 \geq 0.1$
		High-quality coordinated development with synchronized in urban personnel, facility, and management safety resilience	The difference between any two of U_1, U_2, and U_3s consistently less than 0.1

(continued)

Table 3.3 (continued)

Category	Division basis	Concrete type	Specific type classification basis
Good coordinated development	$0.7 \leq D < 0.8$	Good coordinated development with lagging urban personnel safety resilience	$U_2 - U_1 \geq 0.1$ or $U_3 - U_1 \geq 0.1$
		Good coordinated development with lagging urban facility safety resilience	$U_1 - U_2 \geq 0.1$ or $U_3 - U_2 \geq 0.1$
		Good coordinated development with lagging urban management safety resilience	$U_1 - U_3 \geq 0.1$ or $U_2 - U_3 \geq 0.1$
		Good coordinated development with synchronized in urban personnel, facility, and management safety resilience	The difference between any two of U_1, U_2, and U_3s consistently less than 0.1

(continued)

Table 3.3 (continued)

Category	Division basis	Concrete type	Specific type classification basis
Moderately coordinated development	$0.6 \leq D < 0.7$	Moderately coordinated development with lagging urban personnel safety resilience	$U_2 - U_1 \geq 0.1$ or $U_3 - U_1 \geq 0.1$
		Moderately coordinated development with lagging urban facility safety resilience	$U_1 - U_2 \geq 0.1$ or $U_3 - U_2 \geq 0.1$
		Moderately coordinated development with lagging urban management safety resilience	$U_1 - U_3 \geq 0.1$ or $U_2 - U_3 \geq 0.1$
		Moderately coordinated development with synchronized in urban personnel, facility, and management safety resilience	The difference between any two of U_1, U_2, and U_3s consistently less than 0.1

(continued)

Table 3.3 (continued)

Category	Division basis	Concrete type	Specific type classification basis
Primary coordinated development	$0.5 \leq D < 0.6$	Primary coordinated development with lagging urban personnel safety resilience	$U_2 - U_1 \geq 0.1$ or $U_3 - U_1 \geq 0.1$
		Primary coordinated development with lagging urban facility safety resilience	$U_1 - U_2 \geq 0.1$ or $U_3 - U_2 \geq 0.1$
		Primary coordinated development with lagging urban management safety resilience	$U_1 - U_3 \geq 0.1$ or $U_2 - U_3 \geq 0.1$
		Primary coordinated development with synchronized in urban personnel, facility, and management safety resilience	The difference between any two of U_1, U_2, and U_3s consistently less than 0.1

(continued)

Table 3.3 (continued)

Category	Division basis	Concrete type	Specific type classification basis
On the verge of imbalance and decline	$0.4 \leq D < 0.5$	On the verge of imbalance and decline with lagging urban personnel safety resilience	$U_2 - U_1 \geq 0.1$ or $U_3 - U_1 \geq 0.1$
		On the verge of imbalance and decline with lagging urban facility safety resilience	$U_1 - U_2 \geq 0.1$ or $U_3 - U_2 \geq 0.1$
		On the verge of imbalance and decline with lagging urban management safety resilience	$U_1 - U_3 \geq 0.1$ or $U_2 - U_3 \geq 0.1$
		On the verge of imbalance and decline with synchronized in urban personnel, facility, and management safety resilience	The difference between any two of U_1, U_2, and U_3s consistently less than 0.1

(continued)

Table 3.3 (continued)

Category	Division basis	Concrete type	Specific type classification basis
Mild imbalance and decline	$0.3 \leq D < 0.4$	Mild imbalance and decline with lagging urban personnel safety resilience	$U_2 - U_1 \geq 0.1$ or $U_3 - U_1 \geq 0.1$
		Mild imbalance and decline with lagging urban facility safety resilience	$U_1 - U_2 \geq 0.1$ or $U_3 - U_2 \geq 0.1$
		Mild imbalance and decline with lagging urban management safety resilience	$U_1 - U_3 \geq 0.1$ or $U_2 - U_3 \geq 0.1$
		Mild imbalance and decline with synchronized in urban personnel, facility, and management safety resilience	The difference between any two of U_1, U_2, and U_3s consistently less than 0.1

(continued)

Table 3.3 (continued)

Category	Division basis	Concrete type	Specific type classification basis
Moderate imbalance and decline	$0.2 \leq D < 0.3$	Moderate imbalance and decline with lagging urban personnel safety resilience	$U_2 - U_1 \geq 0.1$ or $U_3 - U_1 \geq 0.1$
		Moderate imbalance and decline with lagging urban facility safety resilience	$U_1 - U_2 \geq 0.1$ or $U_3 - U_2 \geq 0.1$
		Moderate imbalance and decline with lagging urban management safety resilience	$U_1 - U_3 \geq 0.1$ or $U_2 - U_3 \geq 0.1$
		Moderate imbalance and decline with synchronized in urban personnel, facility, and management safety resilience	The difference between any two of U_1, U_2, and U_3s consistently less than 0.1
Severe imbalance and decline	$D < 0.2$	Severe imbalance and decline with lagging urban personnel safety resilience	$U_2 - U_1 \geq 0.1$ or $U_3 - U_1 \geq 0.1$
		Severe imbalance and decline with lagging urban facility safety resilience	$U_1 - U_2 \geq 0.1$ or $U_3 - U_2 \geq 0.1$
		Severe imbalance and decline with lagging urban management safety resilience	$U_1 - U_3 \geq 0.1$ or $U_2 - U_3 \geq 0.1$
		Severe imbalance and decline with synchronized in urban personnel, facility, and management safety resilience	The difference between any two of U_1, U_2, and U_3s consistently less than 0.1

Nonparametric Kernel Density Estimation

The kernel density estimation (KDE) is a nonparametric method employing a smoothed curve as a weighting mechanism to estimate the probability density function of sample data (Ye and Jinfeng 2021). This technique is employed to delineate sample data's dynamic and heterogeneous distribution characteristics. It can facilitate more intuitive observation of the temporal evolution of the samples. In contrast to histograms, the KDE generates a more continuous and smooth probability density curve. The function is given as follows:

$$f(x) = \frac{1}{nh}\sum_{i=1}^{n} k\left(\frac{x_i - x}{h}\right) \tag{3.21}$$

where $f(x)$ represents the probability density estimation of urban safety resilience coupling coordination; n represents the number of observations; k symbolizes the kernel function, including Gaussian kernel function, triangular kernel function, and bi-weight kernel function; h is the bandwidth, which is pivotal for the estimation's smoothness and accuracy. It is directly related to curve smoothness and inversely related to estimation precision. Smaller bandwidths are favored for enhanced precision. The primary peak position on the kernel density curve indicates the coupling coordination degree. The variations in peak height and width reflect spatial disparities in coordination among different subsystems. The number of peaks shows the polarization trend in coordination. The spread of the distribution captures the spatial variability in coordination between the most synchronized region and others. A longer tail indicates broader discrepancies (Quah 1993).

Source of Research Area and Indicator Data

1. Overview of the Research Area

The Chengdu-Chongqing Economic Circle is the largest significant metropolitan agglomeration upstream of the Yangtze River. Located at the intersection of the "Belt and Road" and the Yangtze River Economic Belt, it connects the southwest and northwest, links East Asia with Southeast Asia and South Asia. It is the most densely populated, industrially robust, innovative, and open region in western China. In the "Key Tasks of New Urbanization Construction in 2019," the National Development and Reform Commission clearly positioned the Chengdu-Chongqing Urban Agglomeration equally important as the Beijing-Tianjin-Hebei Urban Agglomeration, the Yangtze River Delta Urban Agglomeration, and the Guangdong-Hong Kong-Macao Urban Agglomeration (Shi et al. 2020).

In terms of spatial scope, the total area of the Chengdu-Chongqing Economic Circle reaches 185,000 km^2. It especially encompasses the entire region of

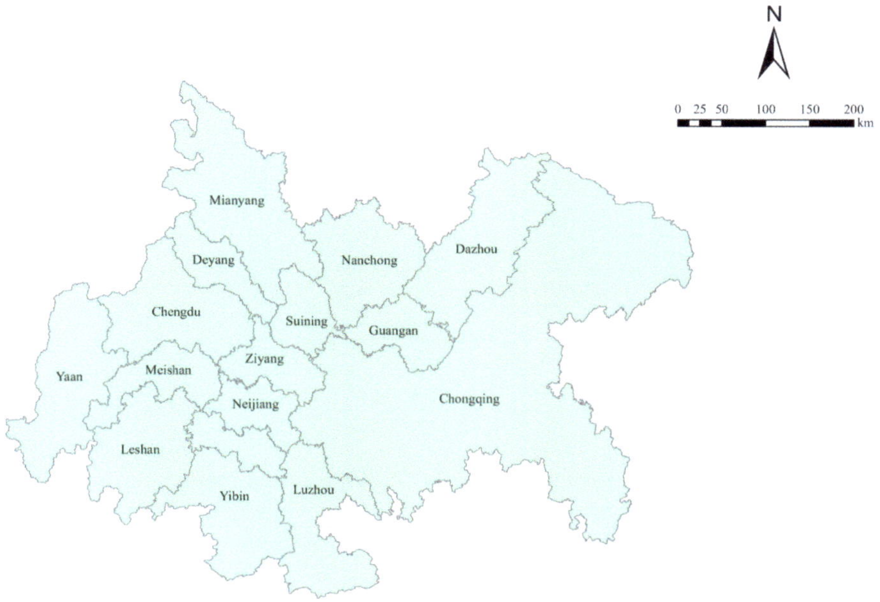

Fig. 3.1 The Chengdu-Chongqing Economic Circle in the Chengyu region

Chongqing and 15 cities in Sichuan Province: Chengdu, Deyang, Mianyang, Meishan, Ziyang, Suining, Leshan, Ya'an, Zigong, Luzhou, Neijiang, Nanchong, Yibin, Dazhou, and Guang'an, for a total of 16 cities (Fig. 3.1).

In terms of economic infrastructure, the statistical data of Sichuan Province and Chongqing City in 2023 show that in 2022, the cities with a GDP of one trillion yuan in the Chengdu-Chongqing region are Chongqing and Chengdu, while the regional GDP of 12 cities such as Mianyang, Yibin, and Deyang is around 100 billion to 400 billion yuan (see Table 3.4 for specific data). By the end of 2023, the GDP of the Chengdu-Chongqing region has reached more than 8 trillion-yuan, accounting for 6.5% of the country and 30.4% of the western region; the GDP increased by 6.1% over the previous year, 0.9% points higher than the national level. It can be seen that the Chengdu-Chongqing Twin Cities Economic Circle is the focus of China's western economic development and a new growth level in China's economic development.

In terms of urbanization development, the urbanization rates of Chongqing and Chengdu reached 70.96% and 79.89%, respectively, in 2022, and the urbanization rates of other cities were around 50% (see Table 3.4 for specific data). The overall urbanization rate of China's 2022 census reached 65.22%, and the overall urbanization rate of the Chengdu-Chongqing region was about 55%. There is still a lot of room for development in regional urbanization.

In terms of the types of public emergencies, they can usually be divided into four categories, as shown in Table 3.5. The Chengdu-Chongqing area needs to deal with many types of public emergencies, and all four types of events are involved. Among

Table 3.4 GDP and urbanization rate of some cities in Chengdu-Chongqing region in 2022

City	GDP (100 million yuan)	Urbanization rate (%)	City	GDP (100 million yuan)	Urbanization rate (%)
Chongqing	29,129.00	70.96	Ya'an	902.51	54.02
Chengdu	20,817.50	79.89	Zigong	1638.42	56.67
Deyang	2816.87	57.61	Luzhou	2601.52	51.91
Mianyang	3626.94	54.29	Neijiang	1656.95	51.56
Meishan	1635.51	51.61	Nanchong	2685.45	51.75
Ziyang	948.16	42.66	Yibin	3427.84	53.55
Suining	1614.47	58.68	Dazhou	2502.72	51.30
Leshan	2308.81	54.57	Guang'an	1425.02	45.32

them, geological disasters occur frequently. In 2022 alone, there were more than 2000 geological disasters of varying sizes in the Chengdu-Chongqing area, and the regional emergency management situation is relatively severe.

2. Indicator data source

The study focuses on the period from 2017 to 2021 for the Chengdu-Chongqing Economic Circle in China, reflecting the relatively recent development of safety and resilience in urban construction in the country. To ensure the timeliness of the data, the sample includes information from 16 cities within the region during this time frame. The primary data sources include "China Statistical Yearbook (2018–2022)," "China Urban Statistical Yearbook (2018–2022)," as well as the statistical yearbooks of individual cities and the statistical bulletins on national economic and

Table 3.5 Common types of public emergencies

Event type	Content
Natural disaster	Geological disasters, meteorological and hydrological disasters, forest fires, biological disasters, etc.
Accident disaster	Road traffic accidents, public facilities and equipment accidents, industrial, mining, commercial and related industry accidents, environmental pollution and ecological damage incidents, etc.
Public health event	Food safety incidents, infectious disease outbreaks, group unexplained diseases, animal epidemics, etc.
Social security incident	Social security incidents, economic security incidents, etc.

social development. In cases where certain indicators lacked data for the year 2018, linear interpolation was applied to estimate the missing values for that specific year. This approach aims to maintain a comprehensive and continuous dataset for the analysis.

Evaluation Measurement and Obstacle Degree Analysis for Safety Resilience Level of Urban Personnel, Facilities, and Management

Weight Result Analysis

Utilizing the acquired data, we shall employ formulas 3.26–3.29 to derive the weights through the entropy method. Simultaneously, in conjunction with expert surveys as detailed in Appendix C, we shall procure the relative importance values of indicators. These shall then be input into the software yaahp10.1 for the determination of weights via the AHP. Finally, by amalgamating the outcomes of both methodologies, we shall obtain the composite weights, with specifics delineated in Table 3.6.

As depicted in Table 3.6, within the urban personnel safety resilience criterion layer, prominent indicators include the number of individuals covered by unemployment insurance, occupational injury insurance, the level of basic medical insurance for urban workers, the population covered by urban pension insurance, and the influx of higher education talent. Notably, the composite weight of security sense and safety culture ranks highest, indicating its paramount contribution to urban personnel safety resilience. Conversely, in the urban facility safety resilience criterion layer, notable indicators encompass the level of urban traffic lighting facilities, urban drainage facilities, the reserve level of personnel in the construction industry, and the penetration rate of cable broadcasting and television. Upon synthesizing the various sub-criterion layers, it becomes apparent that lifeline engineering facilities hold the largest proportion of weight in urban facility safety resilience. Lastly, within the urban management safety resilience criterion layer, noteworthy indicators comprise the comprehensive production capacity of water supply, the fiscal support level for transportation, the level of emergency supplies, and the fiscal support level for public safety. Here, it is discernible that within the sub-criterion layer, the level of material supply holds the highest proportion of weight in urban management safety resilience.

Table 3.6 Final weight of each specific indicator

Serial number	Index	Entropy weight	AHP weight	Comprehensive weight
1	The density of permanent residents in the built-up area	0.006	0.142	0.074
2	The level of basic medical insurance for urban workers	0.148	0.081	0.115
3	The proportion of temporary residents	0.006	0.087	0.046
4	The number of newly added higher education talents	0.112	0.102	0.107
5	The level of reserve health technical personnel in the city	0.113	0.104	0.109
6	The reserve level of urban health institutions	0.056	0.104	0.080
7	The level of social organization units	0.069	0.052	0.061
8	The number of people covered by unemployment insurance	0.166	0.080	0.123
9	The number of people covered by urban old-age insurance	0.115	0.112	0.113
10	The number of people covered by work-related injury insurance	0.154	0.080	0.117
11	The per capita emergency deposit level	0.054	0.057	0.055
12	Urbanization rate	0.059	0.119	0.089
13	Proportion of land area in safety-vulnerable areas	0.029	0.075	0.052
14	The level of personnel reserves in the construction industry	0.172	0.095	0.133
15	Per capita urban road area	0.035	0.059	0.047
16	The level of urban traffic lighting facilities	0.202	0.117	0.159
17	The penetration rate of wired broadcasting and television	0.036	0.165	0.100
18	The level of urban drainage facilities	0.241	0.082	0.162
19	Per capita refuge area	0.065	0.081	0.073
20	Building area of disaster relief reserve institution warehouses per ten thousand people	0.070	0.085	0.077
21	The number of beds in medical and health institutions per ten thousand people	0.064	0.051	0.057
22	The green coverage rate	0.028	0.071	0.050
23	The death rate per million people due to disasters	0.017	0.156	0.087

(continued)

Table 3.6 (continued)

Serial number	Index	Entropy weight	AHP weight	Comprehensive weight
24	The proportion of annual direct economic losses due to disasters to the region's gross domestic product	0.001	0.156	0.079
25	The level of financial support for public safety	0.208	0.062	0.135
26	The level of financial support for medical and health care	0.153	0.039	0.096
27	The level of financial support for transportation	0.184	0.097	0.141
28	The comprehensive production capacity of water supply	0.217	0.196	0.206
29	The level of grain supply	0.021	0.196	0.108
30	The level of emergency material supply	0.198	0.098	0.148

Evaluation Measurement and Obstacle Degree Analysis of Urban Personnel Safety Resilience Level

1. Overall Situation

Based on the weighted data of various indicators in the urban safety resilience coupling coordination evaluation system, calculations were conducted to assess the level of urban personnel safety resilience in the Chengdu-Chongqing Economic Circle from 2017 to 2021. The specific calculation results are presented in Table 3.7 and Fig. 3.2.

Data extracted from Table 3.7 reveals a fluctuating yet ascending trend in the urban personnel safety resilience levels across cities within the Chengdu-Chongqing Economic Circle over the period from 2017 to 2021. Notably, Chengdu and Chongqing consistently lead the economic circle, exhibiting a robust annual enhancement in resilience. Besides these cities, Nanchong and Dazhou occupy the third and fourth positions, respectively, followed closely by Mianyang and Leshan. Zigong and Neijiang also rank highly, while Ziyang, Guang'an, Suining, and Ya'an are at the lower end of the spectrum.

Specifically, both Chengdu and Chongqing have maintained urban personnel safety resilience scores above 0.6 throughout the observed period, with Chengdu surpassing the 0.8 mark in 2020–2021 and Chongqing in 2021. In stark contrast, other cities in the region did not exceed a score of 0.2 during the same timeframe, with the lowest recorded score being 0.066 in Meishan in 2017—this figure is more than tenfold lower than the highest score of 0.881 observed in Chengdu in 2021, highlighting significant regional disparities in urban personnel safety resilience.

While Chengdu, Neijiang, Meishan, Guang'an, Ziyang, and Chongqing showed a consistent upward trajectory each year, other cities experienced some fluctuations

Table 3.7 2017–2021 Chengdu-Chongqing Economic Circle city personnel safety resilience evaluation score

Urban area	2017	2018	2019	2020	2021
Chengdu	0.643	0.692	0.775	0.826	0.881
Zigong	0.135	0.133	0.117	0.144	0.159
Luzhou	0.132	0.129	0.120	0.127	0.130
Deyang	0.120	0.122	0.117	0.130	0.147
Mianyang	0.146	0.151	0.146	0.161	0.162
Suining	0.105	0.101	0.094	0.116	0.106
Neijiang	0.121	0.123	0.137	0.143	0.143
Leshan	0.138	0.147	0.146	0.147	0.145
Nanchong	0.170	0.175	0.160	0.176	0.184
Meishan	0.066	0.122	0.128	0.129	0.131
Yibin	0.121	0.117	0.109	0.112	0.113
Guang'an	0.079	0.083	0.113	0.118	0.125
Dazhou	0.156	0.154	0.165	0.180	0.155
Ya'an	0.115	0.119	0.101	0.109	0.097
Ziyang	0.085	0.096	0.097	0.112	0.116
Chongqing	0.628	0.707	0.721	0.779	0.849

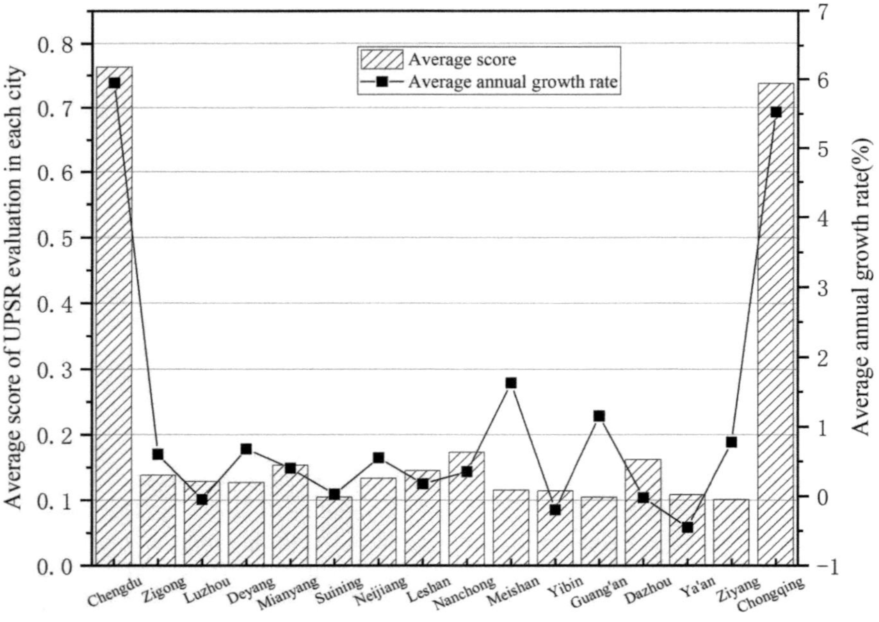

Fig. 3.2 Average score and average annual growth rate of urban personnel safety resilience evaluation in each city

between 2018 and 2021, indicating a general trend of improvement in urban personnel safety resilience levels across the economic circle, albeit with some variances.

Figure 3.2 elucidates that the average urban personnel safety resilience scores for the top five cities are led by Chengdu (0.763) and Chongqing (0.737), followed by Nanchong (0.173), Dazhou (0.162), and Mianyang (0.153). These regions, characterized by higher population densities and advanced personnel safeguarding measures, lead the economic circle. Conversely, the bottom five cities—Yibin (0.114), Ya'an (0.108), Suining (0.104), Guang'an (0.104), and Ziyang (0.101)—display comparatively lower levels of personnel security, attributed to lower population safeguard standards and higher migrant populations, thus contributing to diminished urban personnel safety resilience.

In terms of annual average growth rates, Chengdu and Chongqing are at the forefront with increases of 6.0% and 5.5%, respectively, followed by Meishan and Guang'an, each breaking the 1% growth threshold. In contrast, Luzhou, Yibin, Dazhou, and Ya'an experienced negative growth rates.

2. Diagnostic Assessment of Obstacle Factors

Using an obstacle model, an analysis was conducted on the developmental levels of urban personnel safety resilience across 16 districts within the Chengdu-Chongqing Economic Circle from 2017 to 2021. This involved calculating the obstacle degree for individual indices encompassing basic demographic attributes, social participation readiness, and the sense of safety and cultural subsystems. The aim was to holistically identify the factors impeding the enhancement of urban personnel safety resilience within the economic circle.

(1) Analysis of Obstacle Degrees in Sub-criterion Layer

Data from Table 3.8 illustrates the obstacle degrees for the three major subsystems of urban personnel safety resilience over the specified period. Safety sense and cultural factors emerged as the predominant obstacles, with an average obstacle degree of 46.92%, underscoring them as critical elements in enhancing urban personnel safety resilience. Following this, basic demographic attributes held an average obstacle degree of 33.43%, acting as a connector to the safety sense and culture system, with evident mutual influences noted. The subsystem of social participation readiness presented the lowest obstacle degree at 19.66%, approximately half that of the highest, indicating its relatively minimal impact on the resilience of urban personnel. This can be attributed to the role of social participation as a vital safeguard for personal safety, and the significant national emphasis on developing social organizations and public welfare projects in recent years, which have become crucial factors in population development.

From the perspective of development trends, the obstacle degrees of different sub-criterion layers have different changing trends. The obstacle degrees for basic demographic attributes and social participation readiness subsystems showed an increasing trend, rising from 31.79% and 19.22% in 2017 to 36.65% and 20.21% in 2021, respectively. This indicates an intensifying constraint of these subsystems

Table 3.8 Urban personnel safety resilience sub-criterion level obstacles/%

Year	Basic population attributes	Social participation readiness	Safety perception and safety culture
2017	31.79	19.22	48.99
2018	31.85	19.43	48.73
2019	32.58	19.72	47.69
2020	34.27	19.71	46.03
2021	36.65	20.21	43.14

on urban personnel safety resilience, with demographic attributes showing a particularly notable rise. Conversely, the obstacle degree for the safety sense and culture subsystem showed a general declining trend, decreasing from 48.99% in 2017 to 43.14% in 2021, reflecting ongoing local governmental measures to enhance urban safety culture.

Excluding Chongqing, the obstacle degrees for urban personnel safety resilience were relatively uniform across other regions, primarily hindered by safety sense and culture and basic demographic attributes as detailed in Table 3.9. In 15 cities including Chengdu, the most significant obstacle was identified in the safety sense and culture subsystem, whereas in Chongqing, the primary obstacle resided in the demographic attributes' subsystem. Furthermore, a ranking of the top three obstacle factors for urban personnel safety resilience across the cities showed that for Chongqing, the order was demographic attributes > safety sense and culture > social participation readiness. For the other 15 cities, including Chengdu, Zigong, Luzhou, and Deyang, the sequence was safety sense and culture > demographic attributes > social participation readiness.

(2) Analysis of Obstacle Factors at the Indicator Layer

Based on the subsystem obstacle degree analysis, the indicator level was further scrutinized by focusing on the top five obstacle factors that most significantly constrain urban personnel safety resilience in the Chengdu-Chongqing Economic Circle from 2017 to 2021. The differences in obstacle degrees among these indicators were substantial, ranging from a maximum average obstacle degree of 15.72% to a minimum of 2.59%, a disparity of sixfold.

The primary and fifth-ranked overall obstacle factors were consistently unemployment insurance coverage (P8) and the number of new higher education talents (P4). The second to fourth positions were occupied by urban employee basic medical insurance level (P2), urban old-age insurance coverage (P9), and work injury insurance coverage (P10). Among the top five obstacle indicators, three pertain to the safety sense and culture subsystem, and two derive from the basic demographic attributes' subsystem, with average obstacle degrees of 15.44%, 14.30%, 13.54%, 13.30%, and 12.06%, respectively.

Table 3.9 Mean obstacle degree of urban personnel safety resilience subsystem in municipal areas/%

Urban area	Basic population attributes	Social participation readiness	Safety perception and safety culture
Chengdu	35.69(II)	26.32(III)	37.99(I)
Zigong	29.63(II)	20.83(III)	49.55(I)
Luzhou	32.58(II)	19.30(III)	48.12(I)
Deyang	33.11(II)	19.77(III)	47.11(I)
Mianyang	32.46(II)	19.61(III)	47.94(I)
Suining	32.69(II)	19.28(III)	48.04(I)
Neijiang	30.29(II)	20.04(III)	49.67(I)
Leshan	30.82(II)	20.57(III)	48.61(I)
Nanchong	31.53(II)	18.24(III)	50.23(I)
Meishan	31.77(II)	20.51(III)	47.71(I)
Yibin	33.12(II)	18.86(III)	48.02(I)
Guang'an	32.50(II)	19.69(III)	47.82(I)
Dazhou	28.70(II)	20.17(III)	51.13(I)
Ya'an	30.58(II)	21.06(III)	48.36(I)
Ziyang	31.80(II)	19.81(III)	48.38(I)
Chongqing	57.59(I)	10.43(III)	31.98(II)

Note I, II and III in parentheses indicate the 1st, 2nd, and 3rd ranking of each city's obstacle level

This alignment underscores the predominant obstacles imposed by both the safety sense and culture and the basic demographic attributes subsystems, indicating that the urban personnel safety resilience in the Chengdu-Chongqing Economic Circle is constrained by factors such as unemployment insurance coverage, basic medical insurance levels, and the growth in higher education talent.

To address these constraints, there is a pressing need to further refine the social security systems for unemployment and basic medical insurance, ensuring that citizens receive fundamental life and medical support in times of risk. Additionally, enhancing the quality of higher education and expanding access to premium educational resources are essential to bolster the professional skills and lifelong learning capabilities of urban personnel. Moreover, promoting an equitable distribution of educational resources across urban and rural areas, as well as different regions, will elevate the overall educational standards of the population (Table 3.10).

Table 3.10 Main obstacle factors of urban personnel safety resilience indicator layer obstacle degree/%

Year	Index obstacle ranking				
	OF1	OF2	OF3	OF4	OF5
2017	P8(15.72)	P10(14.98)	P9(13.85)	P2(13.27)	P4(11.97)
2018	P8(15.71)	P10(14.80)	P9(14.01)	P2(12.35)	P4(12.28)
2019	P8(15.58)	P10(14.37)	P9(13.98)	P2(13.33)	P4(11.97)
2020	P8(15.39)	P10(13.96)	P2(13.54)	P9(13.48)	P4(12.50)
2021	P8(14.79)	P2(14.00)	P10(13.37)	P9(12.39)	P4(11.58)

Evaluation Measurement and Obstacle Degree Analysis of Urban Facility Safety Resilience Level

1. Overall Situation

Based on the weighted data of various indicators in the urban safety resilience coupling coordination evaluation system, calculations were conducted to assess the level of urban facility safety resilience in the Chengdu-Chongqing Economic Circle from 2017 to 2021. The specific calculation results are presented in Table 3.11 and Fig. 3.3.

Table 3.11 Chengdu-Chongqing Economic Circle urban facility safety resilience evaluation score in 2017–2021

Urban area	2017	2018	2019	2020	2021
Chengdu	0.494	0.541	0.512	0.552	0.607
Zigong	0.203	0.202	0.218	0.232	0.227
Luzhou	0.197	0.200	0.217	0.236	0.246
Deyang	0.144	0.161	0.207	0.209	0.211
Mianyang	0.206	0.195	0.226	0.248	0.258
Suining	0.215	0.253	0.245	0.265	0.284
Neijiang	0.128	0.152	0.182	0.185	0.212
Leshan	0.174	0.201	0.211	0.216	0.217
Nanchong	0.223	0.251	0.251	0.271	0.292
Meishan	0.197	0.165	0.189	0.179	0.182
Yibin	0.172	0.183	0.230	0.251	0.266
Guang'an	0.196	0.216	0.236	0.240	0.240
Dazhou	0.111	0.173	0.201	0.221	0.211
Ya'an	0.162	0.161	0.190	0.203	0.235
Ziyang	0.128	0.152	0.184	0.183	0.211
Chongqing	0.625	0.748	0.782	0.771	0.799

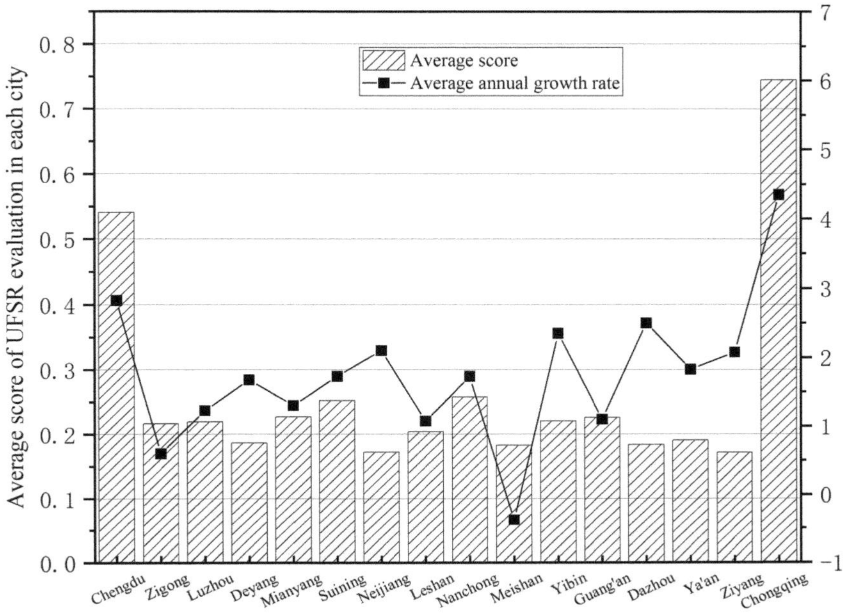

Fig. 3.3 Average score and average annual growth rate of urban facility safety resilience evaluation in each city

Data from Table 3.11 illustrates a fluctuating yet overall ascending trend in the urban facility safety resilience levels across various cities in the Chengdu-Chongqing Economic Circle from 2017 to 2021. Throughout this period, Chongqing consistently maintained a leading position in infrastructure resilience, with Chengdu ranking second. Nanchong and Suining occupied the third and fourth positions, respectively, followed closely by Mianyang and Guang'an. Yibin and Luzhou also ranked relatively high, while Dazhou, Meishan, Neijiang, and Ziyang were at the lower end of the spectrum.

Specifically, Chongqing's infrastructure resilience score consistently exceeded 0.6 over the five-year period, peaking at 0.799 in 2021. Chengdu's scores also surpassed 0.4 throughout the same timeframe, displaying upward trends despite some fluctuations in 2019. In contrast, the remaining cities did not exceed a resilience score of 0.3 during this period, with the lowest score recorded by Dazhou in 2017 at 0.111. This score is more than seven times lower than Chongqing's peak, indicating significant regional disparities in urban facility safety resilience, with Chongqing at the highest and Chengdu following, while other cities displayed closely clustered lower resilience levels. Except for Luzhou, Deyang, Neijiang, Leshan, Nanchong, Yibin, and Guang'an—all of which showed a consistent annual increase in resilience scores from 2017 to 2021—other cities experienced some fluctuations between 2018 and 2021, suggesting a general improvement in urban facility safety resilience levels across the economic circle, albeit with some variability.

As depicted in Fig. 3.3, the average urban facility safety resilience scores rank Chongqing (0.745) and Chengdu (0.541) in the top positions, followed by Nanchong (0.258), Suining (0.252), and Mianyang (0.227). These regions are among the most economically developed and innovative within the entire Chengdu-Chongqing Economic Circle. Conversely, the cities with the lowest rankings—Deyang (0.186), Dazhou (0.183), Meishan (0.182), Neijiang (0.172), and Ziyang (0.172)—exhibit less economic and infrastructural development, which correlates with lower levels of urban facility safety resilience. The highest annual average growth rates were observed in Chongqing (4.4%) and Chengdu (2.8%), with Dazhou, Yibin, Neijiang, and Ziyang also achieving growth rates exceeding 2%, whereas Meishan displayed a negative growth rate.

2. Diagnostic Assessment of Obstacle Factors

Utilizing an obstacle model, the study identified key impediments to the development of urban facility safety resilience across 16 regions within the Chengdu-Chongqing Economic Circle from 2017 to 2021. The model quantified the degree of obstruction for individual indicators across subsystems such as construction engineering, transportation facilities, lifeline infrastructure, and emergency support facilities, aiming to holistically evaluate the factors restraining urban facility resilience within the economic circle.

(1) Analysis of Obstacle Degrees in Sub-criterion Layer

The obstacles of the four major subsystems of urban facility safety resilience in the Chengdu-Chongqing Economic Circle from 2017 to 2021 are shown in Table 3.12. From the perspective of obstacle intensity, construction projects have always been the biggest obstacle to the urban facility safety resilience in the Chengdu-Chongqing Economic Circle, with an average obstacle degree of 33.30%; followed by lifeline engineering facilities, with an average obstacle degree of 32.02%. The difference in average obstacle degrees between lifeline engineering facilities and construction projects is small, but both have a significant impact on the safety and resilience of urban facilities; transportation facilities and emergency support facility subsystems have low obstacle degrees, with averages of 19.26% and 15.42%, respectively, which are the same as those with the highest obstacle degrees. The difference in construction engineering subsystems is nearly one time. It can be seen that transportation facilities and emergency support facilities have relatively little impact on the safety and resilience of urban facilities. The reason is that on the one hand, transportation facilities are relatively developed in the entire Chengdu-Chongqing Economic Circle, and the transportation network is densely arranged; on the other hand, disasters occur frequently in some cities in the Chengdu-Chongqing Economic Circle, and the government attaches great importance to the construction of emergency support facilities.

Table 3.12 Urban facility safety resilience sub-criterion level obstacles/%

Year	Construction engineering	Transportation facilities	Lifeline engineering facilities	Emergency support facilities
2017	32.39	19.61	31.43	16.56
2018	33.15	19.30	31.20	16.35
2019	33.52	19.21	31.86	15.40
2020	33.23	19.40	32.75	14.62
2021	34.22	18.77	32.83	14.18

Examining developmental trends, the obstruction rates for different subsystems in urban facility safety resilience display distinct trajectories. In the Chengdu-Chongqing Economic Circle, both the construction engineering and lifeline infrastructure facilities subsystems show increasing obstruction trends. Their obstruction rates rose from 32.39% and 31.43% in 2017 to 34.22% and 32.83% in 2021, respectively. This increase indicates an intensifying constraint on urban facility safety resilience, particularly in these subsystems, where the escalation is notably pronounced. As urban areas expand, the demand for robust infrastructure to support the growing volume of resources and population density is driving rapid development in the city's lifeline systems. Simultaneously, enhancements in integrated transportation networks are facilitating urban sprawl and population aggregation, further boosting demand in the construction sector.

Conversely, the obstruction rates for the transportation facilities and emergency support subsystems show a general downward trend, decreasing from 19.61% and 16.56% in 2017 to 18.77% and 14.18% in 2021, respectively. Compared to the construction engineering and lifeline infrastructure facilities subsystems, these sectors exhibit relatively lower obstruction rates, and their annual decrease reflects effective measures taken by local governments to improve transportation networks and emergency preparedness.

Analyzing regional variations, it appears that apart from Chongqing, there is not much difference in the other municipal areas. In all cities excluding Chongqing, the principal obstacles to urban facility safety resilience are attributed to construction engineering and lifeline infrastructure facilities. In contrast, Chongqing's primary obstacles stem from construction engineering and emergency support facilities.

When ranking the urban facility safety resilience obstacle factors by city, it emerges that in Chengdu, Zigong, Luzhou, Deyang, and the other 15 cities excluding Chongqing, although the order is not entirely consistent, the top two obstacles consistently pertain to construction engineering and lifeline infrastructure facilities. The last two spots are invariably occupied by transportation facilities and emergency support facilities. Conversely, the sequence of obstacle factors in Chongqing distinctly differs, ranking construction engineering as the most significant, followed by emergency support facilities, lifeline infrastructure facilities, and transportation facilities. This arrangement highlights the unique challenges and priorities faced by Chongqing in enhancing its urban facility safety resilience (Table 3.13).

Table 3.13 Mean obstacle degree of urban facility safety resilience subsystem in municipal areas/%

Urban area	Construction engineering	Transportation facilities	Lifeline engineering facilities	Emergency support facilities
Chengdu	27.93(II)	21.21(IV)	27.93(I)	22.94(III)
Zigong	35.60(I)	19.50(III)	33.52(II)	11.38(IV)
Luzhou	30.97(II)	19.61(III)	35.91(I)	13.51(IV)
Deyang	35.70(I)	18.77(III)	29.03(II)	16.50(IV)
Mianyang	34.76(I)	18.66(III)	32.23(II)	14.34(IV)
Suining	34.13(I)	20.62(III)	29.77(II)	15.48(IV)
Neijiang	34.05(I)	19.19(III)	31.74(II)	15.03(IV)
Leshan	31.82(II)	19.81(III)	32.41(I)	15.96(IV)
Nanchong	35.11(II)	20.74(III)	35.16(I)	8.98(IV)
Meishan	31.14(II)	18.10(III)	34.28(I)	16.49(IV)
Yibin	34.91(II)	19.66(III)	35.30(I)	10.13(IV)
Guang'an	38.20(I)	19.41(III)	31.95(II)	10.44(IV)
Dazhou	30.54(II)	20.17(III)	33.16(I)	16.13(IV)
Ya'an	35.63(I)	18.57(III)	31.12(II)	14.69(IV)
Ziyang	32.92(I)	19.21(III)	32.49(II)	15.37(IV)
Chongqing	29.46(I)	14.89(IV)	26.24(III)	29.42(II)

Note I, II, and III in parentheses indicate the 1st, 2nd, and 3rd ranking of each city's obstacle level

(2) Analysis of Obstacle Factors at the Indicator Layer

Building on the analysis of the sub-criterion layers' obstacle degrees, this section scrutinizes the individual indicators based on their average obstacle degrees to identify the top five most significant obstructive factors, as illustrated in Table 3.14. From 2017 to 2021, the obstacle degrees within the Chengdu-Chongqing Economic Circle's urban facility safety resilience indicator layer varied widely, with the highest average obstacle degree reaching 20.91% and the lowest at 1.93%, marking a tenfold difference.

The third to fifth most significant overall obstacle factors were consistently identified as the level of urban traffic lighting facilities (F5), the prevalence of cable broadcast television (F6), and the urbanization rate (F1). The top two obstacle indicators included the level of employment in the construction sector (F3) and the level of urban drainage facilities (F7). Among the top five obstacle indicators, two are associated with construction engineering and two with lifeline infrastructure facilities, while one stems from the transportation facilities subsystem. The average obstacle degrees for these indicators are 20.66%, 19.78%, 15.99%, 11.36%, and 7.86%, respectively.

Table 3.14 Main obstacle factors of urban facility safety resilience indicator layer obstacle degree/%

Year	Index obstacle ranking				
	OF1	OF2	OF3	OF4	OF5
2017	F7(20.73)	F3(19.13)	F5(16.32)	F6(10.71)	F1(8.31)
2018	F7(20.91)	F3(19.41)	F5(15.88)	F6(10.30)	F1(8.19)
2019	F7(20.57)	F3(19.83)	F5(15.90)	F6(11.29)	F1(7.91)
2020	F7(20.60)	F3(19.83)	F5(16.23)	F6(12.15)	F1(7.53)
2021	F3(20.71)	F7(20.47)	F5(15.60)	F6(12.36)	F1(7.37)

Evaluation Measurement and Obstacle Degree Analysis of Urban Management Safety Resilience Level

1. Overall Situation

Based on the weighted data of various indicators in the urban safety resilience coupling coordination evaluation system, calculations were conducted to assess the level of urban management safety resilience in the Chengdu-Chongqing Economic Circle from 2017 to 2021. The specific calculation results are presented in Table 3.15 and Fig. 3.4.

Table 3.15 Chengdu-Chongqing Economic Circle urban management safety resilience evaluation score in 2017–2021

Urban area	2017	2018	2019	2020	2021
Chengdu	0.459	0.443	0.467	0.521	0.544
Zigong	0.161	0.192	0.180	0.173	0.164
Luzhou	0.171	0.182	0.198	0.189	0.201
Deyang	0.172	0.185	0.183	0.185	0.200
Mianyang	0.216	0.220	0.215	0.221	0.224
Suining	0.175	0.167	0.172	0.168	0.180
Neijiang	0.181	0.183	0.187	0.173	0.168
Leshan	0.165	0.173	0.191	0.178	0.185
Nanchong	0.213	0.216	0.227	0.238	0.242
Meishan	0.136	0.157	0.154	0.155	0.160
Yibin	0.180	0.191	0.209	0.217	0.228
Guang'an	0.174	0.179	0.242	0.237	0.185
Dazhou	0.215	0.139	0.220	0.223	0.228
Ya'an	0.079	0.089	0.117	0.146	0.141
Ziyang	0.168	0.167	0.152	0.165	0.156
Chongqing	0.802	0.818	0.875	0.918	0.949

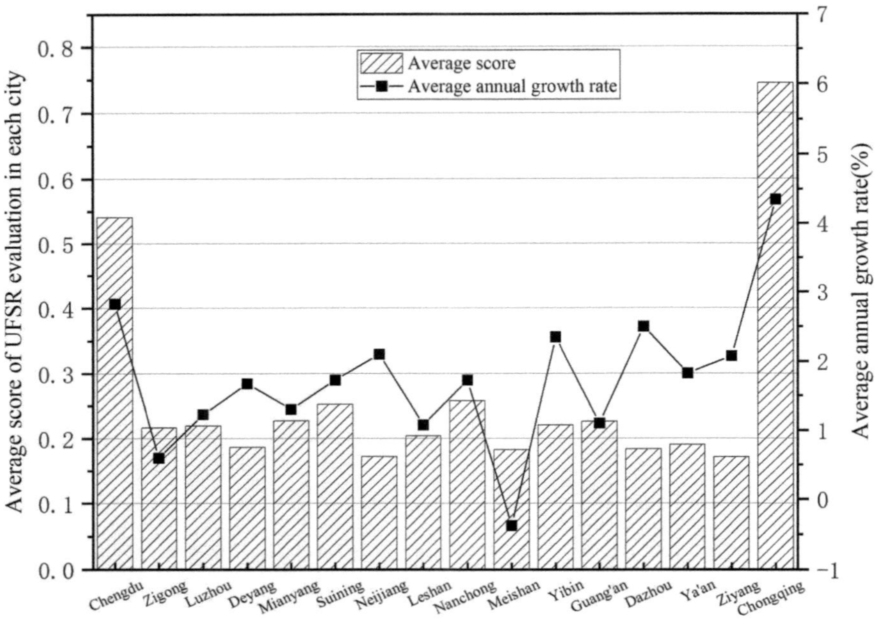

Fig. 3.4 Average score and average annual growth rate of urban management safety resilience evaluation in each city

Table 3.15 reveals that from 2017 to 2021, the urban management safety resilience levels across cities within the Chengdu-Chongqing Economic Circle generally exhibited a fluctuating upward trend. Chongqing consistently led in urban management safety resilience across the economic circle, with Chengdu ranking second, followed by Nanchong and Mianyang in third and fourth places, respectively. Cities like Yibin and Dazhou closely trailed, with Guang'an and Luzhou also ranking high, while Suining, Ziyang, Meishan, and Ya'an were at the bottom.

Specifically, Chongqing's urban management safety resilience scores consistently exceeded 0.8 throughout the period, peaking at 0.949 in 2021. Chengdu's scores were above 0.45, experiencing some fluctuations in 2018 but otherwise showing a rising trend. In contrast, the other cities did not surpass a 0.25 score from 2017 to 2021, with Ya'an recording the lowest at 0.079 in 2017. This was more than 11 times less than Chongqing's highest score of 0.949 in 2021, highlighting substantial regional disparities in urban management safety resilience with Chongqing at the top, followed by Chengdu, while other cities had considerably lower and closely clustered resilience levels.

While the urban management safety resilience of Nanchong, Yibin, and Chongqing consistently improved each year, other cities experienced some fluctuations from 2018 to 2021. Specifically, Zigong, Neijiang, and Guang'an saw continuous declines from 2019 to 2021. This data underscores significant differences in the evolution of urban management safety resilience across the Chengdu-Chongqing Economic Circle, emphasizing the need for targeted improvements in lower-performing cities.

According to Fig. 3.4, the top five cities in terms of average urban management safety resilience scores from 2017 to 2021 are Chongqing (0.872), Chengdu (0.487), Nanchong (0.227), Mianyang (0.219), and Yibin (0.205). These areas are among the more economically developed and have higher levels of disaster prevention and mitigation in the Chengdu-Chongqing Economic Circle. Conversely, the bottom five cities, Zigong (0.174), Suining (0.172), Ziyang (0.162), Meishan (0.152), and Ya'an (0.114), are characterized by less developed economies and lower levels of disaster management, which contribute to their lower urban management safety resilience scores.

In terms of annual average growth rates, Chongqing (3.7%) and Chengdu (2.1%) lead, indicating strong and consistent improvements in urban management safety resilience. Ya'an and Yibin also show significant growth, each with annual increases surpassing 1%. However, Neijiang and Ziyang experienced negative growth rates, suggesting challenges in enhancing their urban management safety resilience amidst other socioeconomic conditions.

2. Diagnostic Assessment of Obstacle Factors

Using the obstacle degree model, the developmental levels of urban management safety resilience in 16 areas of the Chengdu-Chongqing Economic Circle from 2017 to 2021 were analyzed, focusing on the subsystems of risk control level, support guarantee investment, and material supply level. This analysis aims to comprehensively identify the restraining factors impacting the development of urban management safety resilience across the region.

(1) Analysis of Obstacle Degrees in Sub-criterion Layer

As shown in Table 3.16, the level of material supply has consistently been the most significant obstacle to urban management safety resilience in the Chengdu-Chongqing Economic Circle, with an average obstacle degree of 60.68%. Enhancing the level of material supply is identified as the most critical factor for improving urban management safety resilience. Following material supply, support guarantee investment stands as the second most significant challenge, with an average obstacle degree of 35.99%, indicating its substantial impact within the urban management safety resilience system and underscoring the need to enhance this area. Conversely, the obstacle degree for the risk control level subsystem is considerably lower, averaging at 2.55%, which is approximately 20 times less than the most significant obstacle, indicating that risk control is relatively less impactful on urban management safety resilience.

The analysis suggests that the tremendous increase in urban population demands has led to a mismatch in material supply, pointing to the necessity for a more efficient supply chain. Additionally, the level of support guarantee investment in certain cities

Table 3.16 Urban management safety resilience sub-criterion level obstacles/%

Year	Risk control level	Support and guarantee input level	Material supply level
2017	2.00	36.14	61.86
2018	1.99	36.12	61.89
2019	2.01	35.92	62.07
2020	2.48	33.61	60.06
2021	4.28	38.18	57.54

within the Chengdu-Chongqing Economic Circle also affects the orderly progression of disaster prevention and mitigation management.

Examining the trend developments, the obstacle degrees of different subsystems vary. Specifically, the subsystems for risk control level and support guarantee investment have shown an increasing trend in obstacle degrees from 2017 to 2021, rising from 2.00% and 36.14% to 4.28% and 38.18%, respectively. This trend indicates that the constraining effects of both the risk control level and support guarantee investment on urban management safety resilience are intensifying. The increase in the obstacle degree of risk control level is particularly notable, suggesting that there is a critical need to continuously enhance risk control measures within the Chengdu-Chongqing Economic Circle. It is essential to ensure that financial contributions are sufficient to effectively advance disaster prevention and mitigation management.

Conversely, the obstacle degree for the safety perception and cultural system has shown a general downward trend, decreasing from 48.99% in 2017 to 43.14% in 2021. Despite being the highest among the subsystems of population basic attributes and social participation preparation, the annual decrease in the obstacle degree of safety perception and culture reflects ongoing local government measures aimed at enhancing urban safety culture. These efforts are pivotal in mitigating the impacts of existing obstacles and enhancing overall urban management safety resilience.

Analyzing from a regional perspective, there is not much variation among the cities in the Chengdu-Chongqing Economic Circle, with the exception of Chongqing. For the other 15 cities besides Chongqing, the primary obstacles are linked to material supply levels and support guarantee investments, as indicated in Table 3.17. In contrast, in Chongqing, the most significant obstacle affecting urban management safety resilience is the level of material supply.

Upon ranking the factors impeding urban management safety resilience across different cities, it is observed that for cities such as Chengdu, Zigong, Luzhou, and Deyang—excluding Chongqing—the sequence of obstacle factors is consistent, with material supply levels and support guarantee investments being the top two. However, Chongqing differs from the rest; the order of obstacle factors there is material supply levels, followed by risk control levels and then support guarantee investments. This indicates that while most cities in the region are primarily challenged by issues in material supplies and financial support for safety measures, Chongqing additionally faces significant challenges in risk control, underscoring the need for tailored strategies to address these local specificities.

Table 3.17 Mean obstacle degree of urban management safety resilience subsystem in municipal areas/%

Urban area	Risk control level	Support and guarantee input level	Material supply level
Chengdu	1.24(III)	36.89(II)	49.54(I)
Zigong	0.87(III)	36.69(II)	62.43(I)
Luzhou	2.21(III)	36.90(II)	60.89(I)
Deyang	1.08(III)	36.88(II)	62.04(I)
Mianyang	0.61(III)	37.50(II)	61.89(I)
Suining	0.86(III)	36.81(II)	62.34(I)
Neijiang	0.72(III)	36.82(II)	62.46(I)
Leshan	0.89(III)	36.54(II)	62.58(I)
Nanchong	0.75(III)	37.12(II)	62.13(I)
Meishan	1.71(III)	36.22(II)	62.08(I)
Yibin	1.07(III)	36.98(II)	61.96(I)
Guang'an	0.66(III)	38.02(II)	61.32(I)
Dazhou	1.20(III)	37.40(II)	61.40(I)
Ya'an	2.27(III)	35.75(II)	61.98(I)
Ziyang	1.12(III)	36.95(II)	61.93(I)
Chongqing	23.59(II)	22.45(III)	53.96(I)

Note I, II, and III in parentheses indicate the 1st, 2nd and 3rd ranking of each city's obstacle level

(2) Analysis of Obstacle Factors at the Indicator Layer

Building on the analysis of obstacles within the three major subsystems, the top five obstructive factors from 2017 to 2021 within the Chengdu-Chongqing Economic Circle's urban management safety resilience were identified based on the average obstacle degree of individual indicators, as presented in Table 3.18. There was a significant variance in the obstacle degrees of urban management safety resilience indicators across the region, with the maximum average obstacle degree reaching 27.82% and the minimum at 0.30%, a difference of 92 times.

Table 3.18 Main obstacle factors of urban management safety resilience indicator layer Obstacle degree/%

Year	Index obstacle ranking				
	OF1	OF2	OF3	OF4	OF5
2017	M6(27.34)	M8(18.31)	M7(15.21)	M5(14.01)	M3(12.92)
2018	M6(27.82)	M8(18.87)	M7(15.20)	M5(14.22)	M3(12.46)
2019	M6(27.47)	M8(19.09)	M7(15.52)	M5(14.10)	M3(12.53)
2020	M6(25.96)	M8(20.17)	M7(15.71)	M5(14.13)	M3(12.86)
2021	M6(24.74)	M8(17.25)	M5(16.68)	M7(15.55)	M3(12.53)

The primary obstacle factors consistently ranking first, second, and fifth include the comprehensive water supply capacity (M6), emergency material supply level (M8), and public safety financial support level (M3). The third and fourth most significant obstacle indicators are financial support levels for transportation (M5) and food supply levels (M7). Among the top five critical obstacle indicators, three are related to material supply levels and two to support investment inputs. The average obstacle degrees for these indicators are, respectively, 26.67%, 18.94%, 15.44%, 14.63%, and 12.66%.

This distribution underscores the critical areas where improvements are needed, particularly in enhancing material supply chains and financial support mechanisms to bolster urban management safety resilience across the Chengdu-Chongqing economic corridor.

Analysis of Urban Safety Resilience Coupling Coordination Measurement Results

Overall Characteristics of Coupling Coordination Degree

Based on the coupling coordination evaluation results and the coupling coordination degree model, the coupling coordination degree of the Chengdu-Chongqing Economic Circle from 2017 to 2021 was calculated, and the development type classification was completed based on the calculation results. The results are shown in Fig. 3.5 and Table 3.19.

Figure 3.5 illustrates the developmental progression and trends over these years, revealing fluctuations in coupling degree around 0.8, indicative of the urban resilience within the Chengdu-Chongqing Economic Circle oscillating between high and moderately high coupling phases. Meanwhile, the coordination degree of coupling exhibits a yearly ascending trend, escalating from 0.399 in 2017 to 0.436 in 2021, marking a growth rate of 9.27%. Despite this increase, the overall level of coordination remains relatively low. An analysis of both Fig. 3.5 and Table 3.17 permits a characterization of the urban resilience and coordinated development within the Chengdu-Chongqing Economic Circle from 2017 to 2021 as transitioning from a slight disequilibrium and recession type to a near-disequilibrium and recession type, with synchronous development of personnel, facilities, and management resilience. This period can be subdivided into two phases based on the coupling coordination level standards:

(1) "Chengdu-Chongqing Urban Agglomeration" Phase (2017–2019)

Following the national and party strategic plans of 2016, the Chengdu-Chongqing area focused on establishing a "single axis with two belts, dual cores with three zones" structure. This aimed to radiate urban development along the Chengdu-Chongqing axis to surrounding cities and deepen the planning of southern Sichuan, Nansuiguang,

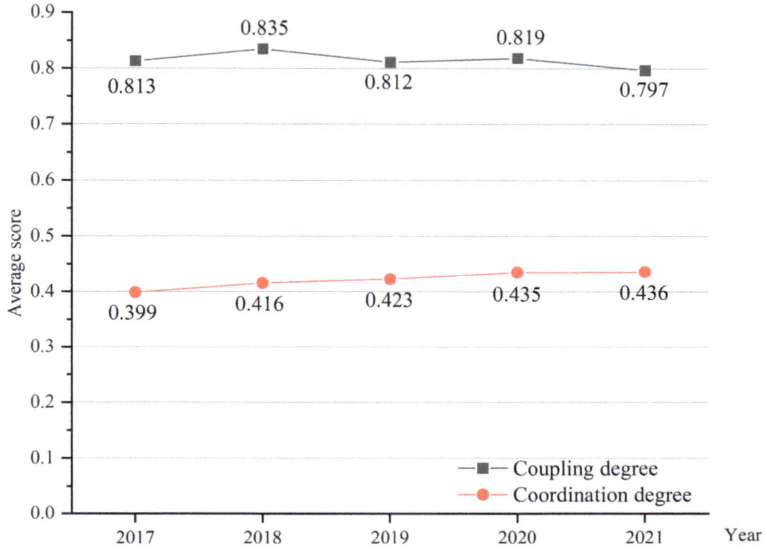

Fig. 3.5 Coupling coordination degree of urban safety resilience in Chengdu-Chongqing Economic Circle from 2017 to 2021

Table 3.19 Classification of urban safety resilience coupling coordination and development type in the Chengdu-Chongqing Economic Circle

Year	Coupling coordination degree	Coupling coordination development type
2017	0.399	Mild imbalance and decline with synchronized in urban personnel, facility, and management safety resilience
2018	0.416	On the verge of imbalance and decline with synchronized in urban personnel, facility, and management safety resilience
2019	0.423	On the verge of imbalance and decline with synchronized in urban personnel, facility, and management safety resilience
2020	0.435	On the verge of imbalance and decline with synchronized in urban personnel, facility, and management safety resilience
2021	0.436	On the verge of imbalance and decline with synchronized in urban personnel, facility, and management safety resilience

and Dawan regions to construct a national-level urban cluster. During this period, regional cooperation and coordinated mechanisms became crucial in emergency management. For instance, Chengdu and Chongqing jointly implemented mechanisms for unified defense against air pollution, ensuring adherence to central government policies. The "Development Plan for Chengdu-Chongqing Urban Agglomeration" highlighted disaster prevention capacity building, underscoring the urgency of disaster prevention and proposed strengthening local government emergency management capabilities and enhancing disaster response capacities to safeguard stable regional economic development.

(2) "Chengdu-Chongqing Economic Circle" Phase (2020–2021)

With the introduction of the "Chengdu-Chongqing Dual-city Economic Circle" in 2020, the development paradigm shifted to "dual-city leadership, dual-circle interaction, and bilateral coordination." In this era, the Chengdu-Chongqing region elevated its developmental goals, emphasizing significant improvements in the socioeconomic development level, enhancing regional vitality, and integration, and cultivating a distinctive Chengdu-Chongqing development model. The "Development Plan for the Chengdu-Chongqing Dual-city Economic Circle" increased the emphasis on economic development within emergency management collaboration, focusing on constructing a modern infrastructure network, intensifying pollution control, promoting public health cooperation, and perfecting emergency coordination mechanisms. This phase accentuated the management of processes, paying close attention to preemptive handling and rescue during disaster events, proposing the sustainability of mechanisms, and emphasizing normalized emergency management practices.

Level of Coupling Coordination Development Across Cities in the Chengdu-Chongqing Economic Circle

Table 3.20 enumerates the measurements of urban safety resilience and coupling coordination for cities within the Chengdu-Chongqing Economic Circle from 2017 to 2021. The data reveal that in 2017, the cities with the highest coupling coordination were Chongqing (0.754), Chengdu (0.655), Nanchong (0.429), Mianyang (0.407), and Leshan (0.382); conversely, the cities with the lowest coordination were Neijiang (0.341), Guang'an (0.334), Ziyang (0.312), Meishan (0.298), and Ya'an (0.297). By 2019, the rankings saw some changes: Chongqing remained the leader, Chengdu retained second place, Nanchong stayed third, Dazhou rose to fourth, and Mianyang dropped to fifth; the cities with the lowest rankings also shifted significantly, listed in descending order as Zigong (0.368), Meishan (0.365), Suining (0.342), Ziyang (0.338), and Ya'an (0.316). By 2021, the cities with the highest coupling coordination continued to be Chongqing (0.874), Chengdu (0.694), and Nanchong (0.443), with Mianyang rising to fourth (0.423) and Dazhou to fifth (0.415); the lowest rankings included Guang'an (0.375), Meishan (0.372), Suining (0.354), Ziyang (0.353), and Ya'an (0.326).

Moreover, examining the average values of urban safety resilience coupling coordination, Chongqing (0.828) leads, followed by Chengdu (0.664) and Nanchong (0.433), with Mianyang (0.413) closely behind. These four cities constitute the "first tier" of coordinated development in urban safety resilience within the Chengdu-Chongqing Economic Circle. Cities like Dazhou (0.399), Leshan (0.391), Luzhou (0.382), Zigong (0.381), Yibin (0.379), Deyang (0.376), Neijiang (0.373), and

Table 3.20 Coupling coordination degree and average of urban safety resilience in cities of the Chengdu-Chongqing Economic Circle, 2017–2021

Urban area	2017	2018	2019	2020	2021	Average
Chengdu	0.655	0.655	0.643	0.675	0.694	0.664
Zigong	0.373	0.391	0.368	0.383	0.389	0.381
Luzhou	0.378	0.382	0.382	0.380	0.387	0.382
Deyang	0.353	0.365	0.371	0.384	0.405	0.376
Mianyang	0.407	0.404	0.411	0.422	0.423	0.413
Suining	0.355	0.345	0.342	0.356	0.354	0.350
Neijiang	0.341	0.360	0.390	0.390	0.382	0.373
Leshan	0.382	0.390	0.398	0.391	0.394	0.391
Nanchong	0.429	0.429	0.424	0.439	0.443	0.433
Meishan	0.298	0.366	0.365	0.367	0.372	0.354
Yibin	0.372	0.375	0.377	0.382	0.387	0.379
Guang'an	0.334	0.338	0.394	0.399	0.375	0.368
Dazhou	0.348	0.376	0.418	0.440	0.415	0.399
Ya'an	0.297	0.310	0.316	0.342	0.326	0.318
Ziyang	0.312	0.339	0.338	0.359	0.353	0.340
Chongqing	0.754	0.829	0.834	0.847	0.874	0.828

Guang'an (0.368) with average coupling coordination ranging between 0.36 and 0.4 form the "second tier." Conversely, Suining (0.350), Ziyang (0.340), and Ya'an (0.318) with averages below 0.36 belong to the "third tier."

In summary, there is a significant disparity in the level of urban safety resilience and coupling coordination among the cities of the Chengdu-Chongqing Economic Circle. Chongqing and Chengdu significantly outperform other cities in terms of coupling coordination, while Meishan, Suining, Ziyang, and Ya'an continually rank lower, indicating a spatial lock-in effect. This suggests that regions with more developed economies tend to have higher urban safety resilience and coupling coordination, whereas less economically advanced cities exhibit lower levels.

Urban Safety Resilience Coupling Coordination Types

To enhance the clarity of observations regarding the coordinated development types and characteristics of the three subsystems of urban safety resilience, an analysis of panel data from 2017 to 2021 was conducted. The 16 cities within the Chengdu-Chongqing Economic Circle can be classified into four types of urban safety resilience coupling coordination: cities with lagging personnel safety resilience, cities with lagging facilities safety resilience, cities with lagging management safety resilience, and cities with synchronous development in personnel, facilities, and management safety resilience (Table 3.21). From a broader perspective, the coupling coordination level of the Chengdu-Chongqing Economic Circle has gradually improved, ascending from mild disequilibrium in 2017 to approaching disequilibrium in 2021, with only Chongqing achieving good and excellent coordination levels. The remaining cities were predominantly at the mild disequilibrium level, indicating significant potential for improvement in regional coupling coordination.

In 2017, the predominant category among the 16 cities was mild disequilibrium with synchronous development in personnel, facilities, and management safety resilience, encompassing 7 cities (43.75% of the study area), including Zigong, Luzhou, Deyang, Neijiang, Leshan, Yibin, and Ziyang. Following this, the cities categorized under mild disequilibrium with lagging personnel safety resilience and approaching disequilibrium with synchronous development each constituted 12.50% of the circle. Cities like Suining and Guang'an fell under the former category, showing weaker personnel safety resilience compared to facilities and management. Mianyang and Nanchong, included in the latter category, displayed smaller disparities among the three resilience subsystems. Other categories such as medium coordination with lagging facilities and management, moderate disequilibrium with lagging personnel, and good coordination with lagging personnel and facilities resilience were less common, each accounting for 6.25%.

By 2018, Meishan transitioned from moderate disequilibrium with lagging personnel safety resilience to mild disequilibrium with synchronous development; Dazhou shifted from mild disequilibrium with lagging facilities safety resilience to mild disequilibrium with synchronous development; and Ya'an moved from moderate to mild disequilibrium with synchronous development. Chongqing progressed from good coordination with lagging personnel and facilities safety resilience to excellent coordination with lagging personnel safety resilience. These changes highlight significant economic disparities within the Chengdu-Chongqing Economic Circle, with Chongqing, as a municipality directly under the central government, boasting a robust economic foundation, infrastructure, and management capabilities compared to other cities.

By 2019, although the prevalent category remained cities with mild disequilibrium and synchronous development, their number decreased from 10 to 7, comprising 43.75% of the circle. Cities like Zigong and Yibin shifted to mild disequilibrium with lagging personnel safety resilience, increasing its proportion to 25.00%; Dazhou escalated from mild to approaching disequilibrium. Overall, the cities within the circle showed an upward trend in coupling coordination levels, yet the development of personnel safety resilience remained uneven, warranting attention.

In 2020, there were no changes in the coupling coordination levels among the 16 cities, though some cities experienced shifts in their coupling types. Zigong, for example, moved from mild disequilibrium with lagging personnel safety resilience to synchronous development, while Luzhou shifted from synchronous development to lagging personnel safety resilience. Chongqing progressed from excellent coordination with lagging personnel safety resilience to excellent coordination with lagging

Table 3.21 Coupling coordination types of urban safety resilience across cities in the Chengdu-Chongqing Economic Circle

Urban area	2017	Type	2018	Type	2019	Type	2020	Type	2021	Type
Chengdu	0.655	Moderately coordinated development with lagging urban facility safety resilience and urban management safety resilience	0.655	Moderately coordinated development with lagging urban facility safety resilience and urban management safety resilience	0.643	Moderately coordinated development with lagging urban facility safety resilience and urban management safety resilience	0.675	Moderately coordinated development with lagging urban facility safety resilience and urban management safety resilience	0.694	Moderately coordinated development with lagging urban facility safety resilience and urban management safety resilience
Zigong	0.373	Mild imbalance and decline with synchronized in urban personnel, facility, and management safety resilience	0.391	Mild imbalance and decline with synchronized in urban personnel, facility, and management safety resilience	0.368	Mild imbalance and decline with lagging urban personnel safety resilience	0.383	Mild imbalance and decline with synchronized in urban personnel, facility, and management safety resilience	0.389	Mild imbalance and decline with synchronized in urban personnel, facility, and management safety resilience
Luzhou	0.378	Mild imbalance and decline with synchronized in urban personnel, facility, and management safety resilience	0.382	Mild imbalance and decline with synchronized in urban personnel, facility, and management safety resilience	0.382	Mild imbalance and decline with synchronized in urban personnel, facility, and management safety resilience	0.380	Mild imbalance and decline with lagging urban personnel safety resilience	0.387	Mild imbalance and decline with lagging urban personnel safety resilience

(continued)

Table 3.21 (continued)

Urban area	2017	Type	2018	Type	2019	Type	2020	Type	2021	Type
Deyang	0.353	Mild imbalance and decline with synchronized in urban personnel, facility, and management safety resilience	0.365	Mild imbalance and decline with synchronized in urban personnel, facility, and management safety resilience	0.371	Mild imbalance and decline with synchronized in urban personnel, facility, and management safety resilience	0.384	Mild imbalance and decline with synchronized in urban personnel, facility, and management safety resilience	0.405	On the verge of imbalance and decline with synchronized in urban personnel, facility, and management safety resilience
Mianyang	0.407	On the verge of imbalance and decline with synchronized in urban personnel, facility, and management safety resilience	0.404	On the verge of imbalance and decline with synchronized in urban personnel, facility, and management safety resilience	0.411	On the verge of imbalance and decline with synchronized in urban personnel, facility, and management safety resilience	0.422	On the verge of imbalance and decline with synchronized in urban personnel, facility, and management safety resilience	0.423	On the verge of imbalance and decline with synchronized in urban personnel, facility, and management safety resilience
Suining	0.355	Mild imbalance and decline with lagging urban personnel safety resilience	0.345	Mild imbalance and decline with lagging urban personnel safety resilience	0.342	Mild imbalance and decline with lagging urban personnel safety resilience	0.356	Mild imbalance and decline with lagging urban personnel safety resilience	0.354	Mild imbalance and decline with lagging urban personnel safety resilience and urban management safety resilience

(continued)

Table 3.21 (continued)

Urban area	2017	Type	2018	Type	2019	Type	2020	Type	2021	Type
Neijiang	0.341	Mild imbalance and decline with synchronized in urban personnel, facility, and management safety resilience	0.360	Mild imbalance and decline with synchronized in urban personnel, facility, and management safety resilience	0.390	Mild imbalance and decline with synchronized in urban personnel, facility, and management safety resilience	0.390	Mild imbalance and decline with synchronized in urban personnel, facility, and management safety resilience	0.382	Mild imbalance and decline with synchronized in urban personnel, facility, and management safety resilience
Leshan	0.382	Mild imbalance and decline with synchronized in urban personnel, facility, and management safety resilience	0.390	Mild imbalance and decline with synchronized in urban personnel, facility, and management safety resilience	0.398	Mild imbalance and decline with synchronized in urban personnel, facility, and management safety resilience	0.391	Mild imbalance and decline with synchronized in urban personnel, facility, and management safety resilience	0.394	Mild imbalance and decline with synchronized in urban personnel, facility, and management safety resilience
Nanchong	0.429	On the verge of imbalance and decline with synchronized in urban personnel, facility, and management safety resilience	0.429	On the verge of imbalance and decline with synchronized in urban personnel, facility, and management safety resilience	0.424	On the verge of imbalance and decline with synchronized in urban personnel, facility, and management safety resilience	0.439	On the verge of imbalance and decline with synchronized in urban personnel, facility, and management safety resilience	0.443	On the verge of imbalance and decline with lagging urban personnel safety resilience

(continued)

Table 3.21 (continued)

Urban area	2017	Type	2018	Type	2019	Type	2020	Type	2021	Type
Meishan	0.298	Moderate imbalance and decline with lagging urban personnel safety resilience	0.366	Mild imbalance and decline with synchronized in urban personnel, facility, and management safety resilience	0.365	Mild imbalance and decline with synchronized in urban personnel, facility, and management safety resilience	0.367	Mild imbalance and decline with synchronized in urban personnel, facility, and management safety resilience	0.372	Mild imbalance and decline with synchronized in urban personnel, facility, and management safety resilience
Yibin	0.372	Mild imbalance and decline with synchronized in urban personnel, facility, and management safety resilience	0.375	Mild imbalance and decline with synchronized in urban personnel, facility, and management safety resilience	0.377	Mild imbalance and decline with lagging urban personnel safety resilience	0.382	Mild imbalance and decline with lagging urban personnel safety resilience	0.387	Mild imbalance and decline with lagging urban personnel safety resilience
Guang'an	0.334	Mild imbalance and decline with lagging urban personnel safety resilience	0.338	Mild imbalance and decline with lagging urban personnel safety resilience	0.394	Mild imbalance and decline with lagging urban personnel safety resilience	0.399	Mild imbalance and decline with lagging urban personnel safety resilience	0.375	Mild imbalance and decline with lagging urban personnel safety resilience
Dazhou	0.348	Mild imbalance and decline with lagging urban personnel safety resilience	0.376	Mild imbalance and decline with synchronized in urban personnel, facility, and management safety resilience	0.418	On the verge of imbalance and decline with synchronized in urban personnel, facility, and management safety resilience	0.440	On the verge of imbalance and decline with synchronized in urban personnel, facility, and management safety resilience	0.415	On the verge of imbalance and decline with synchronized in urban personnel, facility, and management safety resilience

(continued)

Table 3.21 (continued)

Urban area	2017	Type	2018	Type	2019	Type	2020	Type	2021	Type
Ya'an	0.297	Moderate imbalance and decline with synchronized in urban personnel, facility, and management safety resilience	0.310	Mild imbalance and decline with synchronized in urban personnel, facility, and management safety resilience	0.316	Mild imbalance and decline with synchronized in urban personnel, facility, and management safety resilience	0.342	Mild imbalance and decline with synchronized in urban personnel, facility, and management safety resilience	0.326	Mild imbalance and decline with lagging urban personnel safety resilience
Ziyang	0.312	Mild imbalance and decline with synchronized in urban personnel, facility, and management safety resilience	0.339	Mild imbalance and decline with synchronized in urban personnel, facility, and management safety resilience	0.338	Mild imbalance and decline with synchronized in urban personnel, facility, and management safety resilience	0.359	Mild imbalance and decline with synchronized in urban personnel, facility, and management safety resilience	0.353	Mild imbalance and decline with synchronized in urban personnel, facility, and management safety resilience
Chongqing	0.754	Good coordinated development with lagging urban personnel safety resilience and urban facility safety resilience	0.829	High-quality coordinated development with lagging urban personnel safety resilience	0.834	High-quality coordinated development with lagging urban personnel safety resilience	0.847	High-quality coordinated development with lagging urban personnel safety resilience and urban facility safety resilience	0.874	High-quality coordinated development with lagging urban personnel safety resilience and urban facility safety resilience

personnel and facilities safety resilience. These changes likely stemmed from imbalances in the development of urban safety resilience subsystems, where some cities excelled in certain areas but lagged in others.

In 2021, only Deyang exhibited a change in coupling coordination level, while other cities saw shifts in their coupling types. For instance, Suining moved from mild disequilibrium with lagging personnel safety resilience to mild disequilibrium with lagging personnel and management safety resilience; Nanchong transitioned from approaching disequilibrium with synchronous development to approaching disequilibrium with lagging personnel safety resilience; and Ya'an shifted from synchronous development to lagging personnel safety resilience. These changes could be attributed to imbalances in resource allocation, including financial support, technical assistance, and human resources, among the various subsystems of urban safety resilience.

In conclusion, although the coupling coordination levels within the Chengdu-Chongqing Economic Circle are progressively improving, most cities remain at a mild disequilibrium level, indicating substantial room for enhancement in urban safety resilience development across the region. Factors such as uneven economic development, uncoordinated growth of urban safety resilience subsystems, unbalanced resource distribution, and mismatches between policy implementation and local needs are evident. It is advisable to promote information exchange and resource sharing within the circle, establish tighter cooperation mechanisms to address common issues in urban safety resilience development, and through policy guidance and financial support, foster balanced growth within the circle, particularly supporting cities lagging in infrastructure and management systems. These measures can promote balanced development of urban safety resilience within the Chengdu-Chongqing Economic Circle, enhancing the region's overall capacity and resilience in facing natural disasters and societal challenges.

Temporal Evolution of Urban Safety Resilience Coupling Coordination

In this study, kernel density estimation was employed to analyze the coupling coordination level of urban safety resilience, examining changes in its distribution shape, location, and extensiveness. Using MATLAB 2022, kernel density estimation curves of coupling coordination were plotted to explore the dynamic evolution of urban safety resilience coupling coordination across cities within the Chengdu-Chongqing Economic Circle.

Figure 3.6, depicting the kernel density estimation curve of urban safety resilience coupling coordination for the Chengdu-Chongqing Economic Circle, describes the dynamic evolution of the distribution trends in urban safety resilience coupling coordination throughout the region. Initially, the main peak of the kernel density curve exhibits no significant rightward shift over the period, suggesting a modest improvement in urban safety resilience coupling coordination over time, consistent with

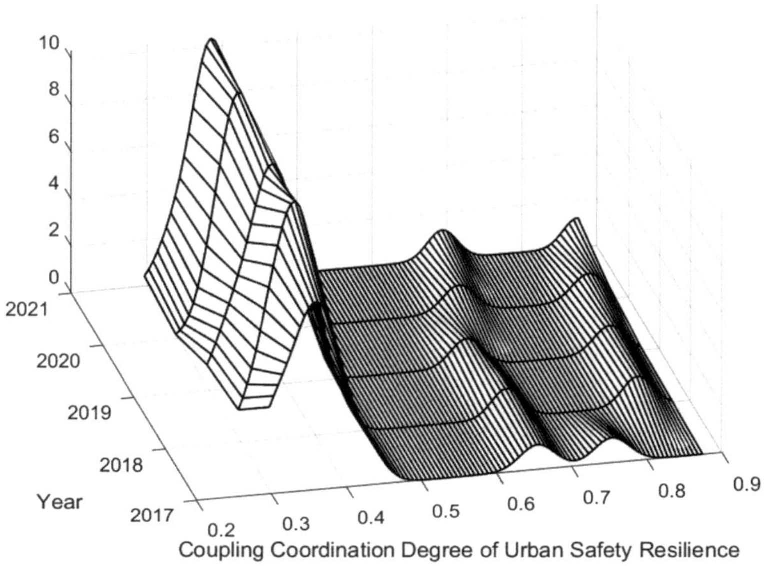

Fig. 3.6 Kernel density curve of coupling coordination degree of urban safety resilience

previous quantitative measurements. Additionally, the kernel density function is characterized by a prominent main peak and two lower side peaks, indicating a notable polarization in the coupling coordination levels among cities within the Chengdu-Chongqing Economic Circle. This polarization reveals that while most cities cluster within a specific range of coupling coordination levels, only a few cities exhibit exceptionally high levels of coordination.

Lastly, the distribution curve of coupling coordination levels shows a bulge on the right side during the study period, without exhibiting a converging trend. This implies that the gap between regions with high and low levels of coupling coordination has not narrowed over time. Such findings highlight the persistent disparities in urban safety resilience and suggest a need for targeted interventions to enhance the coordination level in cities lagging behind, to foster a more balanced regional development in terms of urban safety and resilience.

References

Ariken, M., F. Zhang, K. Liu, et al. 2020. Coupling coordination analysis of urbanization and eco-environment in Yanqi Basin based on multi-source remote sensing data. *Ecological Indicators* 114: 106331.

Biao, Q., and F. Chuanglin. 2012. The dynamic coupling model of the harmonious development between urbanization and eco-environment and its application in arid area. *Acta Ecologica Sinica* 25 (11): 3003–3009 (in Chinese).

Cheng, Z., F. Xuegang, and T. Rui. 2016. Analysis and forecast of coupling coordination development among the regional economy-ecological environment-tourism industry—A case study of provinces along the Yangtze Economic Zone. *Economic Geography* 36 (3): 186–193 (in Chinese).

Chuansong, Z., R. Jianlan, C. Yanbin, and L. Kai. 2018. A study of coupled coordination and spatial and temporal differentiation between science and technology innovation and sustainable development in China. *Scientia Geographica Sinica* 38 (2): 214–222 (in Chinese).

Dasgupta, S., B. Laplante, H. Wang, et al. 2002. Confronting the environmental Kuznets Curve. *Journal of Economic Perspectives* 16 (1): 147–168.

Deyou, M., L. Yuqi, F. Xinsheng, and S. Benlin. 2013. Coordination degree between transportation and regional economic development based on projection pursuit model in Henan. *Geographical Research* 32 (11): 2092–2106 (in Chinese).

Fawei, L., X. Hengzhou, and W. Shu. 2014. Analysis on spatial-temporal coupling coordinate degree among population, land and economy urbanization: Based on China provincial panel data. *Urban Development Studies* 21 (8): 7–11 (in Chinese).

Gang, W., Z. Huabing, X. Fei, and Z. Yan. 2017. Relations between land use carbon budget and economic development at county level in Chengdu city. *Journal of Natural Resources* 32 (7): 1170–1182 (in Chinese).

Honghao, Z., and C. Xiaohong. 2019. Coupling relationship and its response mechanism between smart development and environmental quality in Chinese resource-based cities. *Journal of Natural Resources* 34 (8): 1620–1632 (in Chinese).

Jie, Z., L. Tongsheng, and W. Wuke. 2010. Quantitative analysis of coupling status of man-land relationship areal system in Weihe River Basin. *Progress in Geography* 29 (6): 733–739 (in Chinese).

Jilai, L., L. Yansui, L. Yurui, and H. Yingen. 2018. Coupling analysis of rural residential land and rural population in China during 2007–2015. *Journal of Natural Resources* 33 (11): 1861–1871 (in Chinese).

Jinchuan, H., and F. Chuanglin. 2003. Analysis of coupling mechanism and rules between urbanization and eco-environment. *Geographical Research* 22 (2): 211–220 (in Chinese).

Lei, Z., P. Tingting, F. Cheng, and L. Shuang. 2020. Tourism industry and new urbanization—A view based on system coupling perspective. *Tourism Tribune* 35 (1): 14–31 (in Chinese).

Li, M., L. Hualou, Z. Yingnan, T. Shuangshuang, and G. Dazhuan. 2018. Spatio-temporal coupling relationship between agricultural labor changes and agricultural economic development at county level in China and its implications for rural revitalization. *Acta Geographica Sinica* 73 (12): 2364–2377 (in Chinese).

Lijun, C., Z. Dajie, and X. Yuying. 2000. Study on the framework of an integrated regional sustainable development model. *Journal of Quantitative & Technological Economics* 4: 64–67 (in Chinese).

Ma, S., Z. Li, L. Li, et al. 2022. Coupling coordination degree spatiotemporal characteristics and driving factors between new urbanization and construction industry: Evidence from China. *Engineering, Construction and Architectural Management*.

Mei, G., Q. Bing, and Z. Xiuxia. 2021. The evolution of the spatiotemporal pattern of the coupling and coordination between economic growth kinetic energy conversion and green development. *Geographical Research* 40 (9): 2572–2590 (in Chinese).

Panpan, C., Z. Yuan, X. Siyou, and Y. Jiyao. 2020. Characteristics of ecological environment and high-quality development measurement and spatial and temporal coupling in the Yellow River Basin. *Economic Geography* 40 (5): 49–57+80 (in Chinese).

Quah, D. 1993. Galton's fallacy and tests of the convergence hypothesis. *The Scandinavian Journal of Economics* 95 (4): 427–443.

Shi, Y., X. Cao, D. Shi, et al. 2020. The "one-city monopoly index": Measurement and empirical analysis of China. *Cities* 96: 102434.

Shujia, W., K. Wei, R. Liang, Z. Dandan, and D. Binting. 2021. Research on misuses and modification of coupling coordination degree model in China. *Journal of Natural Resources* 36 (3): 793–810 (in Chinese).

Wei, G., and L. Yongfeng. 2012. The analysis of spatiotemporal evolution of economic and environmental coordination development degree in Liaoning coastal economic belt. *Geographical Research* 31 (11): 2044–2054 (in Chinese).

Weibing, T., and T. Tanling. 2017. Study on the coupling effect between new urbanization development and financial support in the central provinces' cities. *China Soft Science* 3: 140–151 (in Chinese).

Xiangyang, G., M. Xueqing, D. Zhengshan, and M. Qingzhong. 2021. The coordination pattern of tourism efficiency and high-speed transportation: A case study of 41 cities in the Yangtze River Delta. *Geographical Research* 40 (4): 1042–1063 (in Chinese).

Xiaomin, G., F. Chuanglin, M. Xufang, et al. 2022. Coupling and coordination analysis of urbanization and ecosystem service value in Beijing-Tianjin-Hebei urban agglomeration. *Ecological Indicators* 137: 108782.

Xiaonan, C. 2019. Expression and mathematical property of coupling model, and its misuse in geographical science. *Economic Geography* 39 (4): 18–25 (in Chinese).

Xiaoqin, M. 2019. Measurement of the coupling degree of rural tourism and rural revitalization in Shanxi province. *Chinese Journal of Agricultural Resources and Regional Planning* 40 (9): 257–262 (in Chinese).

Xisheng, H., H. Wei, and W. Chengzhen. 2013. Analysis of the coupling of land ecosystem services and urbanization in Fuzhou City. *Scientia Geographica Sinica* 33 (10): 1216–1223 (in Chinese).

Xuemei, Y., Y. Taibao, S. Peiji, W. Wenjie, and L. Haimang. 2014. Coupling effect of water resources-urbanization composite system in north west arid region of China: A case of Shiyang River Basin. *Arid Land Geography* 37 (1): 19–30 (in Chinese).

Yan, Z., and W. Yuming. 2006. Analysis of coupling mechanism and space-time coordination of industrialization and urbanization in regions of China. *Urban Development Studies* 6: 46–51 (in Chinese).

Yang, C., W. Zeng, and X. Yang. 2020a. Coupling coordination evaluation and sustainable development pattern of geo-ecological environment and urbanization in Chongqing municipality, China. *Sustainable Cities and Society* 61: 102271.

Yang, Y., W. Bao, and Y. Liu. 2020b. Coupling coordination analysis of rural production-living-ecological space in the Beijing-Tianjin-Hebei region. *Ecological Indicators* 117: 106512.

Yansui, L., and L. Yurui. 2010. Spatio-temporal coupling relationship between farmland and agricultural labor changes at county level in China. *Acta Geographica Sinica* 65 (12): 1602–1612 (in Chinese).

Yaobin, L., L. Rendong, and S. Xuefeng. 2005. Analysis of the coupling between urbanization and ecological environment in China. *Journal of Natural Resources* 1: 105–112 (in Chinese).

Ye, X., and Z. Jinfeng. 2021. Measurement of the coupling degree of innovation elements allocation and economic high-quality development in China. *Journal of Quantitative & Technological Economics* 38 (10): 46–64 (in Chinese).

Yunfei, L., L. Hongmei, and M. Hongyang. 2021. Research on the evaluation of China's agricultural modernization level-based on entropy method and TOPSIS method. *Agricultural Economic Issues* 2: 107–116 (in Chinese).

Zhenjie, S., Y. Jiadong, and L. Jibin. 2020. Evaluation and obstacle factors of life services level of typical tourism cities in China. *Scientia Geographica Sinica* 40 (2): 289–297 (in Chinese).

Chapter 4
Spatial Analysis of Urban Safety Resilience Coupling Coordination Degree in the Chengdu-Chongqing Economic Circle

This chapter employs GIS and spatial modeling technologies to comprehensively analyze the spatial dynamics of urban safety resilience coupling coordination within the Chengdu-Chongqing Economic Circle. It outlines the process of constructing a spatial model that integrates urban safety resilience data from 2017 to 2021 with geographic information to examine the spatial distribution and evolution patterns across the sixteen cities in the region.

The findings indicate that the spatial distribution in the Chengdu-Chongqing area consistently highlights Chengdu and Chongqing as regions of high value, while other cities exhibit varying degrees of safety resilience. The region maintains a high level of resilience overall, with a dual-core dominance in its economic circle. The variations in other cities reflect changes in economic and development policies, infrastructure improvements, and investment levels. This analysis is crucial for understanding the interplay of urban safety resilience and spatial dynamics in the Chengdu-Chongqing Economic Circle, providing insights into regional development and strategic planning.

Spatial Evolution Analysis of Urban Safety Resilience Coupling Coordination Degree

Methodology and Construction of the Spatial Model

1. Layer Theory in Spatial Analysis

Spatial analysis constitutes an integral aspect of geographic information systems (GIS) that involves organizing thematic data into map layers based on their geometric properties and attributes. The concept of layer theory in spatial analysis was first

Y. Xiang et al., *Research on Coupling Coordination of Urban Safety Resilience Based on Disaster Prevention and Reduction*, Resilient and Friendly Living Environments,
https://doi.org/10.1007/978-981-96-1562-9_4

introduced by Philip Levis in 1962. Levis introduced the methodological innovation of evaluating natural landscape resources as a distinct layer, allowing for a comprehensive understanding of each element. The strategic superposition of these layers unveiled comprehensive characteristics, laying the groundwork for what would evolve into the established layer theory in GIS. The data layers are typically classified into two categories: vector and raster layers. Each layer type enables different spatial analysis methods, namely vector spatial analysis and raster spatial analysis.

(1) Overlay Analysis

Overlay analysis is a technique that involves conducting a sequence of operations on datasets within the same spatial reference system. This process combines new data that encapsulates the thematic content of existing map layers. The emergent data amalgamates the attributes of the original layers, revealing the relationships between spatial features and exclusive attributes. The method is instrumental in uncovering variations, connections, and transformations across multiple data layers. Common techniques employed in overlay analysis include point and polygon overlay, line and polygon overlay, and polygon overlay, each producing distinct data. Advanced GIS software assigns each feature specific attribute, improving the accuracy and relevance of research findings.

(2) Comparative Analysis

Comparative analysis within GIS is pivotal for identifying differences across spatial entities, supporting integrated analytical efforts. The analysis is divided into two parts: data comparative analysis and vector comparative analysis. The former is utilized to discern distinctions in spatial data, enabling change detection, automatic updates, and mitigating errors caused by data inaccuracies. Vector comparative analysis involves analyzing the attributes and characteristics of spatial entities, allowing for a comprehensive comparison of vector data from diverse sources. This approach is critical for identifying and scrutinizing disparities in data.

(3) Network Analysis

Network analysis in GIS focuses on analyzing network elements' spatial and attribute data by considering their topological relationships. It involves using mathematical models to analyze network performance characteristics. The primary objective of network analysis is to optimize network configurations to maximize operational efficiency, which is commonly done in different production, social, and economic domains. Network analysis facilitates the creation of network datasets for urban infrastructure such as electricity networks, water supply systems, and communication networks. It also utilizes GIS software for geographical analysis and modeling, contributing to these domains' research.

2. Construction of the Spatial Model

Before developing a spatial model for the coordination of urban safety resilience coupling across the Chengdu-Chongqing Economic Circle, it is essential to obtain various geographical information from the National Geomatics Center of China. The ArcGIS software is employed to create geographical information data at a scale of 1:200,000 using the latest standard maps provided by the Chinese Ministry of Natural Resources. From this data, relevant geographical information for the 16 regions within the Chengdu-Chongqing Economic Circle is selected and set as new layers. The data on urban safety resilience coupling coordination from 2017 to 2021 for the 16 cities within the circle are connected to the corresponding city layers in ArcGIS. It creates a GIS spatial model that evaluates urban safety resilience coupling coordination across these regions during the specified time period.

Overall Goal Evaluation Level Spatial Distribution

After establishing the GIS spatial model for evaluating the urban safety resilience coupling coordination of the 16 regions within the Chengdu-Chongqing Economic Circle from 2017 to 2021, the Jenks Natural Breaks tool in ArcGIS is used to categorize the urban safety resilience coupling coordination from 2017 to 2021 into four zones: high-value areas, sub-high value areas, sub-low-value areas, and low-value areas. The classification creates spatial distribution maps of urban safety resilience coupling coordination for these regions, visually illustrating city-level coupling coordination's spatial differentiation and evolution patterns. The maps are presented explicitly in Figs. 4.1, 4.2, 4.3, 4.4 and 4.5.

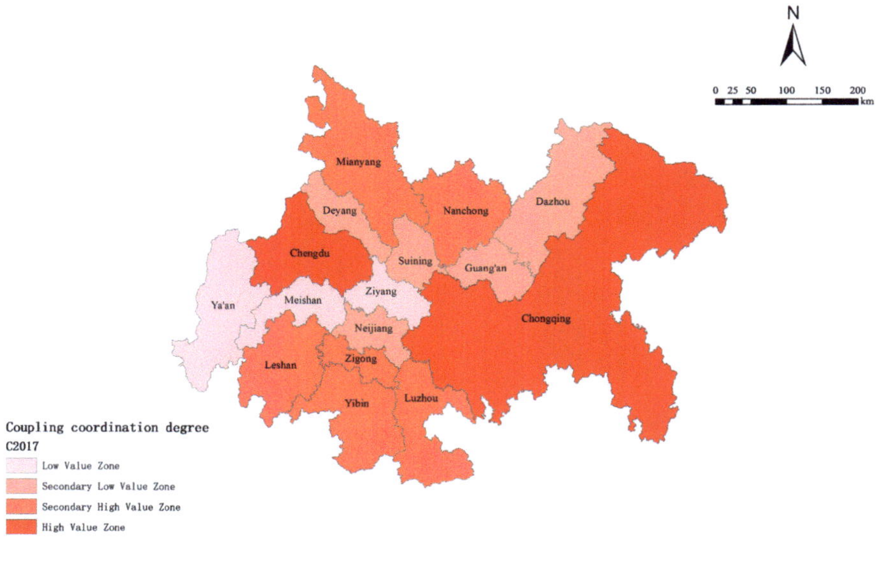

Fig. 4.1 Spatial pattern of coupling coordination degree of urban safety resilience in 2017

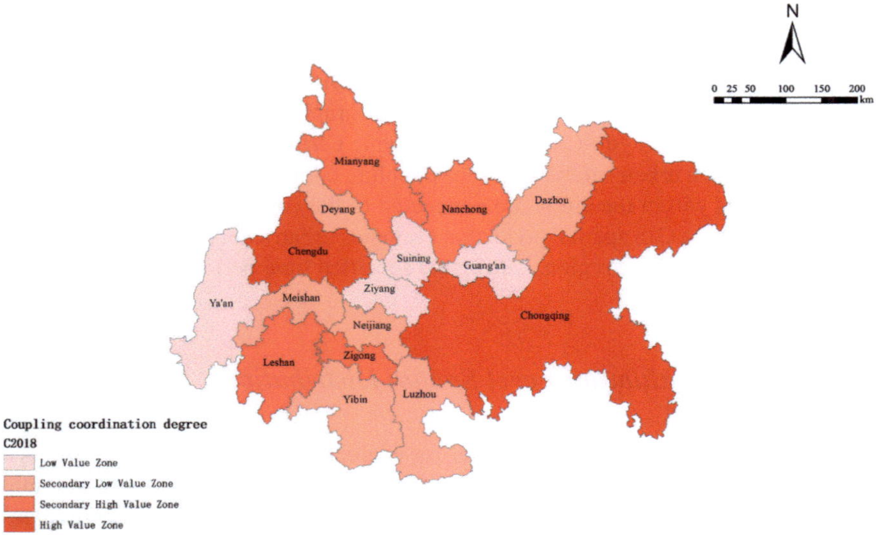

Fig. 4.2 Spatial pattern of coupling coordination degree of urban safety resilience in 2018

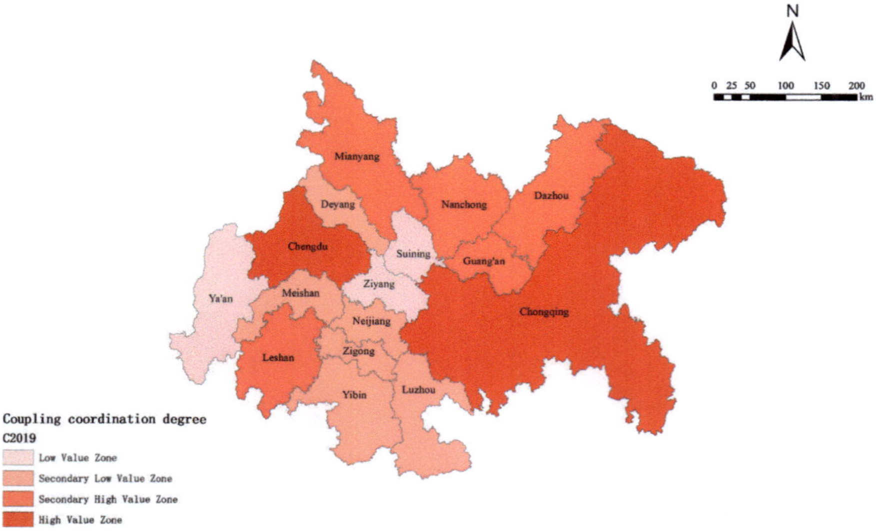

Fig. 4.3 Spatial pattern of coupling coordination degree of urban safety resilience in 2019

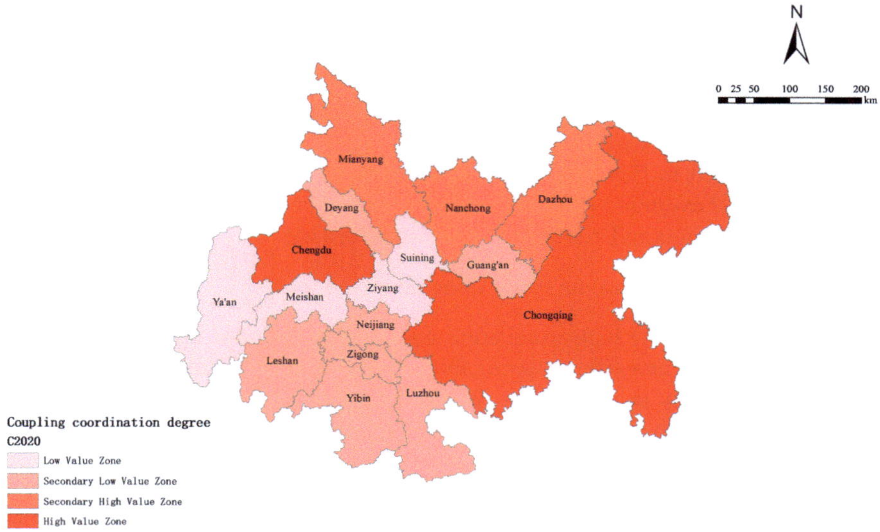

Fig. 4.4 Spatial pattern of coupling coordination degree of urban safety resilience in 2020

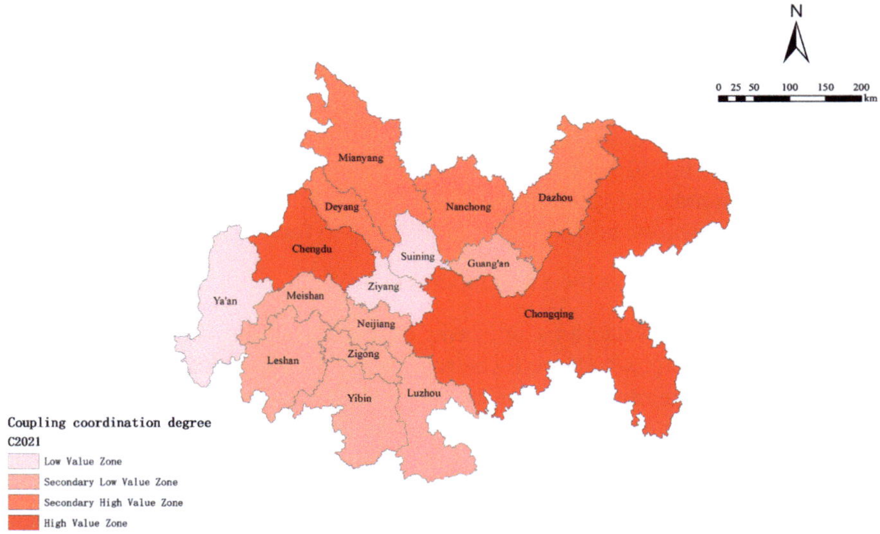

Fig. 4.5 Spatial pattern of coupling coordination degree of urban safety resilience in 2021

These figures visually depict the differences and trends in urban safety resilience across the economic circle and offer critical insights into areas requiring targeted improvements and further research. This analysis serves as a foundation for strategic planning and intervention to enhance urban resilience and safety across the Chengdu-Chongqing Economic Circle.

As can be seen in Figs. 4.1, 4.2, 4.3, 4.4 and 4.5, several critical observations about the spatial evolution of urban safety resilience coupling coordination within the Chengdu-Chongqing Economic Circle can be drawn:

(1) Spatial Distribution of the High-Value Areas

From 2017 to 2021, the high-value areas consistently included only Chengdu and Chongqing, with no new cities entering this category. Chengdu and Chongqing have performed superiorly in urban safety resilience coupling coordination, surpassing other cities. These two cities occupy an irreplaceable core position within the economic circle.

(2) Spatial Distribution of the Sub-High-Value Areas

The allocation of sub-high-value areas reflects the dynamic changes in regional development. In 2017, this category included six cities: Mianyang, Nanchong, Leshan, Zigong, Yibin, and Luzhou. These cities are primarily located in the northwestern and southern parts of the circle. In 2018, Luzhou and Yibin dropped to the sub-low-value areas, reducing the number of cities in the sub-high category to four. In 2019, Zigong dropped to the sub-low-value area, while Guang'an and Dazhou moved up from the low and sub-low-value areas. As a result, the total number in this category increased to five. In 2020, Leshan and Guang'an dropped to sub-low-value areas, decreasing the number of sub-high-value areas to three. In 2021, Deyang moved up to the sub-high category, bringing the total back up to four.

(3) Spatial Distribution of the Sub-Low-Value Area

In 2017, the sub-low-value area primarily included Dazhou, Guang'an, Suining, Deyang, and Neijiang, which were centered in the northern part of the economic circle. In 2018, Suining and Guang'an experienced a decline in coupling coordination degree, moving them into the low-value area. Meishan, Yibin, and Luzhou dropped from the sub-high to the sub-low-value area, increasing the number of cities in this category to six. In 2019, Zigong also fell to the sub-low-value area, while Dazhou moved up, and the focus shifted from the north to the south. In 2020, the downgrade of Guang'an and Leshan increased the number of cities in the sub-low category to seven. In 2021, Deyang's upgrade to sub-high and Meishan's downgrade to low shifted the distribution again.

(4) Spatial Distribution of the Low-Value Areas

In 2017, the low-value areas included Ziyang, Meishan, and Ya'an, primarily located in the central-western part. In 2018, Meishan upgraded to the sub-low-value area, and Suining and Guang'an downgraded to the low-value area, expanding the number to four cities. In 2019, Guang'an moved up to the sub-high area. In 2020, Meishan returned to the low-value area. In 2021, Meishan upgraded again, reducing the low-value area to three cities.

Overall, Chengdu and Chongqing city are the only areas that belong to high-value areas of coupling coordination within the Chengdu-Chongqing Economic Circle that consistently exhibit significant levels of coupling coordination. The other cities exhibit significant developmental imbalances and dynamic changes, presenting a "dual-core dominance" distribution pattern. Chengdu and Chongqing hold unmistakably central roles within the circle, with a marked gap in urban safety resilience and coupling coordination compared to other cities. This disparity stems from a complex array of factors, including levels of economic development, infrastructure, policy support, and geographic location.

Chengdu and Chongqing significantly surpass other cities in urban safety resilience coupling coordination. This is because Chengdu is the capital city of Sichuan province, and Chongqing is a municipality in Southwestern China. These cities have more economic capabilities and more comprehensive infrastructures, enabling them to manage various risks more effectively. They receive significant support from both central and local governments, enjoy numerous policy benefits and investments, and occupy strategic geographic positions serving regional hubs for transportation and logistics, fostering economic concentration and growth. Promoting cooperation between Chengdu, Chongqing, and other cities and sharing resources and information can enhance the resilience of the entire economic circle. Additionally, disseminating the successful experiences of Chengdu and Chongqing in urban planning and risk management to other cities could elevate the overall coupling coordination across the region.

The variation in the number of cities within the sub-high-value areas reflects internal developmental imbalances and competition within the region. Evaluations of the developmental levels of various municipal subsystems reveal that these cities adopt different development strategies based on their resources and advantages, leading to variations in urban safety resilience coupling coordination. Macroeconomic fluctuations and changes in local policies impact cities' economic development and safety resilience, while disparities in infrastructure investment among cities result in varying capacities to manage risks.

To tackle these problems, a strategy focused on balanced development and increased investment is recommended. Implementing regional balanced growth strategies can narrow the developmental gaps between cities. Increasing investment in infrastructure and key sectors can boost cities' overall resilience and competitiveness. The changing numbers and geographic distributions of cities in the sub-low and low-value areas reflect the lag in developing some cities and the lack of stable mechanisms to enhance urban safety resilience in others. For instance, resource and funding allocations are more concentrated in high-value area cities, leading to developmental shortages in other cities; some cities struggle to enhance their safety resilience and coordination due to weak economic foundations; and the lack of effective inter-city cooperative mechanisms within the region leads to insufficient resource sharing and risk-sharing.

Therefore, establishing tighter regional cooperation mechanisms to promote resource sharing and collaborative development is crucial. Strengthening infrastructure construction and improvement in low and sub-low-value areas will enhance their attractiveness and competitiveness. Most critically, customized development strategies tailored to the specific circumstances of each city are necessary to address these disparities and promote balanced regional development effectively.

Spatial Correlation Analysis of Urban Safety Resilience Coupling Coordination

Methodology and Construction of the Spatial Matrix

1. Quantitative Geography in Spatial Analysis

Quantitative geography is a discipline that employs mathematical methods and computer technology to establish geographical models and quantitatively analyze relationships between geographical elements. It can simulate the spatiotemporal evolution of geographical systems, providing a scientific basis for optimizing human–environment interactions. As quantitative geography develops, it has merged with mathematical models, computational methods, and 3S technology, evolving into a methodological system that uses computer technology to handle complex data and conduct comprehensive research. Modern applications of spatial statistical analysis in quantitative geography are extensive. It involves establishing statistical connections between data based on their spatial positions and extracting spatial dependencies, associations, and autocorrelations related to geographical locations from the data. This process assists in supporting the optimization of human–environment relationships.

Classical statistical analysis methods assume sample independence. However, certain spatial studies may reveal spatial connections and associations within urban spatial data that cannot be directly revealed using traditional statistical approaches, mainly because geographical locations influence these connections. Therefore, modifications are necessary on top of classical statistical techniques to make them suitable for the statistical analysis of spatial data.

(1) Spatial Weight Matrix

The spatial weight matrix is an effective way to express spatial relationships. It quantitatively represents the "spatial structure" among data and is a prerequisite and foundation for exploratory spatial data analysis (ESDA).

The spatial weight matrix W, describing the spatial dependence relationships among n entities, is denoted as W_{ij}. W_{ij} represents the degree of influence of entity i on entity j. It can be measured based on proximity criteria or distance standards.

$$W = \begin{bmatrix} W_{11} & W_{12} & \cdots & W_{1n} \\ W_{21} & W_{22} & \cdots & W_{2n} \\ \vdots & \vdots & & \vdots \\ W_{n1} & W_{n2} & \cdots & W_{nn} \end{bmatrix} \tag{4.1}$$

There are two main common rules for determining the spatial weight matrix:

① Adjacency matrix

$$\omega_{ij} = \begin{cases} 1 \text{ When regions } i \text{ and } j \text{ are adjacent} \\ 0 \text{ Other} \end{cases} \tag{4.2}$$

② Distance-based spatial weights

$$\omega_{ij} = \begin{cases} 1 \text{ When the distance between region } i \text{ and } j \text{ is less than distance} \\ 0 \text{ Other} \end{cases} \tag{4.3}$$

(2) Global Spatial Autocorrelation

Moran's I and Geary's C are two global indicators used to measure spatial autocorrelation. Moran's I reflects the similarity of attribute values for spatially adjacent or neighboring units, while Geary's C negatively correlates with the Moran Index.

If x_i is the observed value for location (or region) i, the global Moran's I can be calculated using the following:

$$\boldsymbol{I} = \frac{n\sum_{i=1}^{n}\sum_{j=1}^{n} \boldsymbol{w}_{ij}(\boldsymbol{x}_i - \overline{\boldsymbol{x}})(\boldsymbol{x}_j - \overline{\boldsymbol{x}})}{\sum_{i=1}^{n}\sum_{j=1}^{n} \boldsymbol{w}_{ij}\sum_{i=1}^{n}(\boldsymbol{x}_i - \overline{\boldsymbol{x}})^2} = \frac{\sum_{i=1}^{n}\sum_{j=1}^{n} \boldsymbol{w}_{ij}(\boldsymbol{x}_i - \overline{\boldsymbol{x}})(\boldsymbol{x}_j - \overline{\boldsymbol{x}})}{\boldsymbol{S}^2\sum_{i=1}^{n}\sum_{j=1}^{n} \boldsymbol{w}_{ij}} \tag{4.4}$$

where I represents the Moran's I; $S^2 = \frac{1}{n}\sum_{i=1}^{n}(x_i - \overline{x})^2$; $\overline{x} = \frac{1}{n}\sum_{i=1}^{n} x_i$.

Then Geary's C can be calculated as follows:

$$\boldsymbol{C} = \frac{(n-1)\sum_{i=1}^{n}\sum_{j=1}^{n} \boldsymbol{w}_{ij}(\boldsymbol{x}_i - \overline{\boldsymbol{x}})^2}{2\sum_{i=1}^{n}\sum_{j=1}^{n} \boldsymbol{w}_{ij}\sum_{i=1}^{n}(\boldsymbol{x}_i - \overline{\boldsymbol{x}})^2} \tag{4.5}$$

where C is Geary's C; other variables are the same as (4.3).

The value of Moran's I is generally between − 1 and 1. A value less than 0 indicates a negative correlation, a value equal to 0 indicates no correlation, and a value greater than 0 indicates a positive correlation. The value of Geary's C is generally between 0 and 2. A value greater than 1 indicates that a negative correlation, a value equal to 1 shows no correlation, a value less than 1 indicates a positive correlation.

For Moran's I, the standardized statistic Z is commonly used to test whether there is a spatial autocorrelation relationship in n regions. The calculation formula of Z is:

$$Z = \frac{I - E(I)}{\sqrt{\text{VAR}(I)}} \tag{4.6}$$

In practice, it is also necessary to perform significance tests on spatial autocorrelation relationships. Here are some statistical test used to determine the significance of Moran's I statistic:

① Normal Distribution Assumption

Considering an infinitely large sample size, the Z-value conforms to a standard normal distribution. The significance level can be determined as this.

② Randomization Method

This method assumes that the observation values and their locations within the regional units are entirely unrelated, implying that the Z-value asymptotically follows a standard normal distribution. This assumption is used to determine the significance level.

③ Radom Permutation Test

Radom permutation test assumes that observation values can appear with equal probability at any location. However, the distribution of Moran's I is empirically generated by randomly reordering observation values across all spatial units, calculating different Moran's I with each iteration, and the mean and variance of Moran's I can be obtained.

A positive and significant Z-value indicates positive spatial autocorrelation, meaning that similar observation values tend to cluster together spatially whether they are high or low. Conversely, a negative and significant Z-value indicates negative spatial autocorrelation, implying that similar observation values tend to be dispersed. A Z-value of zero indicates that the observation values are independently and randomly distributed.

(3) Local Spatial Autocorrelation

When evaluating spatial autocorrelation on a global scale using indices Moran's I and Geary's C, it may not capture the inherent variability in spatial processes. When the research aims to identify local clusters with either high or low values, determine which regional units most significantly influence global spatial autocorrelation, and assess the extent to which global metrics conceal localized anomalies or micro-scale instabilities, analyzing local spatial autocorrelation becomes crucial. Various local spatial autocorrelation analysis methodologies include local indicators of spatial association (LISA), G statistics, and the Moran scatterplot, each tailored to different analytical needs.

① Local Indicators of Spatial Association (LISA)

The LISA was developed by Luc Anselin in 1995. It can identify spatial clusters and outliers. Each LISA statistic is associated with a single spatial unit, allowing it to assess the degree of significant spatial clustering of similar or dissimilar values around that unit.

② G Statistics

The G statistic is a local spatial autocorrelation measure that identifies hot and cold spots in spatial data. It operates by examining the intensity of clustering of within a specified distance range.

③ Moran Scatterplot

The Moran scatter plot is commonly employed to examine the volatility of local spatial patterns. It presents a graphical depiction of the spatial lag factor in two dimensions. The Moran scatter plot is divided into four quadrants, each denoting a different local spatial connection between an urban unit and its neighbors. In the first quadrant (High–High, HH), there is a spatial connection where locations with high values are surrounded by neighbors with high values. In the second quadrant (Low–High, LH), there is a spatial connection where locations with low values are surrounded by neighbors with high values. In the third quadrant (Low–Low, LL), there is a spatial connection where locations with low values are surrounded by neighbors with low values. In the fourth quadrant (High–Low, HL), there is a spatial connection where locations with high values are surrounded by neighbors with low values (Yong-jie and Zhong-ying 2011).

Integrating the Moran scatterplot with the LISA significance levels can generate a Moran significance map. This map highlights significant LISA regions and identifies corresponding areas based on the different quadrants of the Moran scatterplot. Given a confidence level, if I_i is significant > 0 and $z_i > 0$, the region i is located in the first quadrant; if I_i is significant > 0 and $z_i < 0$, the region i is located in the third quadrant; if I_i is significant < 0 and $z_i > 0$, the region i is located in the second quadrant; if I_i is significant < 0 and $z_i < 0$, the region i is located in the fourth quadrant (Yingxia et al. 2005).

This study's model constructed in ArcGIS primarily addresses the relationship between the geographical spaces of the 16 cities within the Chengdu-Chongqing Economic Circle and their urban safety resilience coupling coordination. Consequently, this book selects the Rook Contiguity model, which can reflect the relevant spatial distribution relationships, fulfilling the spatial matrix requirements for the regions. The approach has led to generalizing a spatial adjacency matrix detailing these relationships.

Global Spatial Correlation of Urban Safety Resilience Coupling Coordination

The global spatial correlation is primarily evaluated using Moran's I and its corresponding z-score. The spatial correlation tools were used available in ArcGIS to calculate Moran's. Considering the distance can influence the principal order, the Euclidean distance has been chosen to analyze global spatial correlations (Rulian et al. 2021). Utilizing the spatial autocorrelation tool within ArcGIS, an analysis was conducted on the urban safety resilience coupling coordination of the different cities. The calculations of Moran's I and other relevant indices have been conducted, with specific data presented in Figs. 4.6, 4.7, 4.8, 4.9 and 4.10.

As shown in Figs. 4.6, 4.7, 4.8, 4.9 and 4.10, Moran's I of the spatial distribution of urban safety resilience coupling coordination were − 0.064, − 0.034, 0.018, 0.020, and 0.009, respectively. All values within the range of − 1 to 1 are not equal to zero. It indicates that spatial clustering of urban safety resilience coupling coordination is possible. However, these values are close to zero, indicating that the degree of clustering is very low.

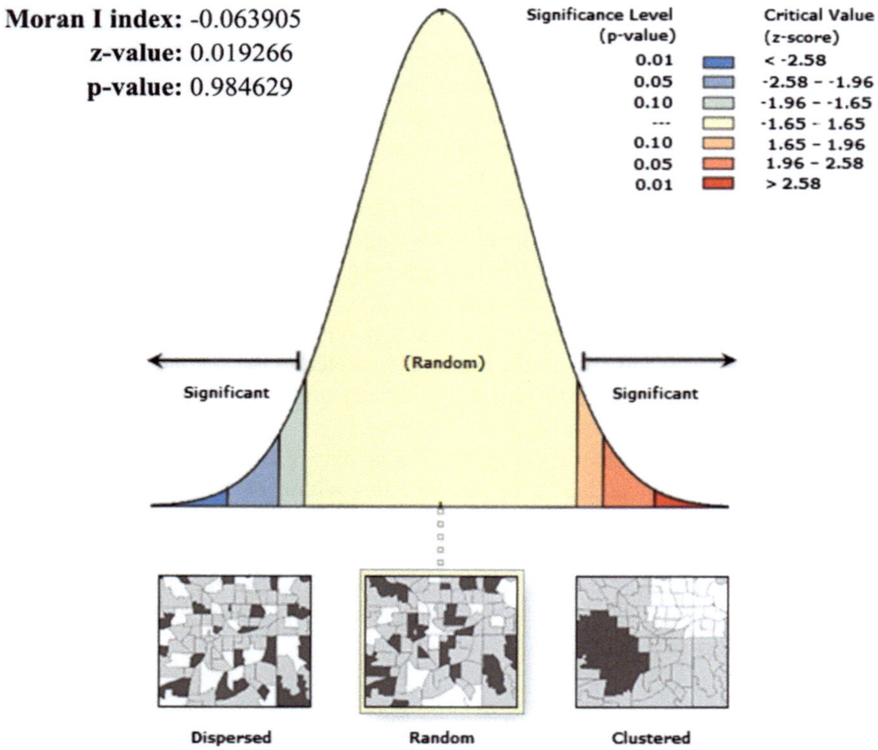

Fig. 4.6 Results of the spatial correlation analysis in 2017

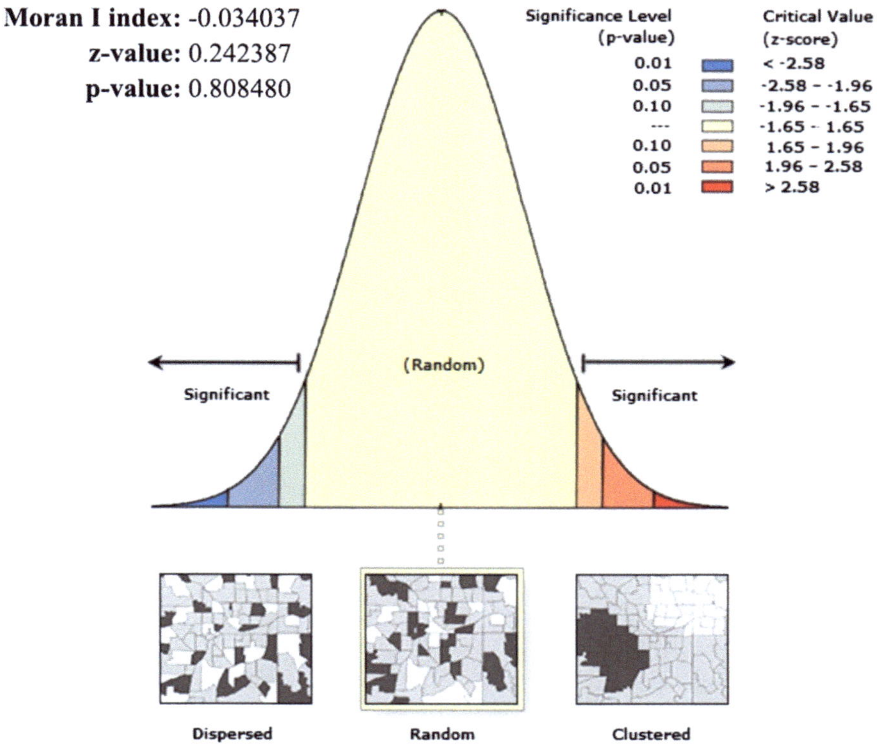

Fig. 4.7 Results of the spatial correlation analysis in 2018

The corresponding *p*-values for these indices were 0.985, 0.808, 0.523, 0.520, and 0.573, respectively, reflecting the probability that the spatial data could have been generated randomly. These probabilities indicate a likelihood greater than that of a random distribution, but they do not significantly reject the null hypothesis. The *z*-scores from 2017 to 2021 were 0.019, 0.242, 0.639, 0.643, and 0.563, respectively. It indicates that there is no significant spatial correlation or clustering potential (Zhigang et al. 2006).

The disparity in economic development levels among cities within the Chengdu-Chongqing Economic Circle is likely a pivotal factor contributing to the low spatial autocorrelation observed. The construction of safety resilience in different cities is constrained by their economic conditions, leading to randomness in their spatial distribution. The level of urban infrastructure development directly influences their safety resilience coupling coordination. The uneven infrastructure development within the economic circle may contribute to variations in urban safety resilience among cities, thereby affecting spatial autocorrelation.

Therefore, implementing unified policy standards for enhancing urban safety resilience is imperative. Ensuring all cities within the circle can progress

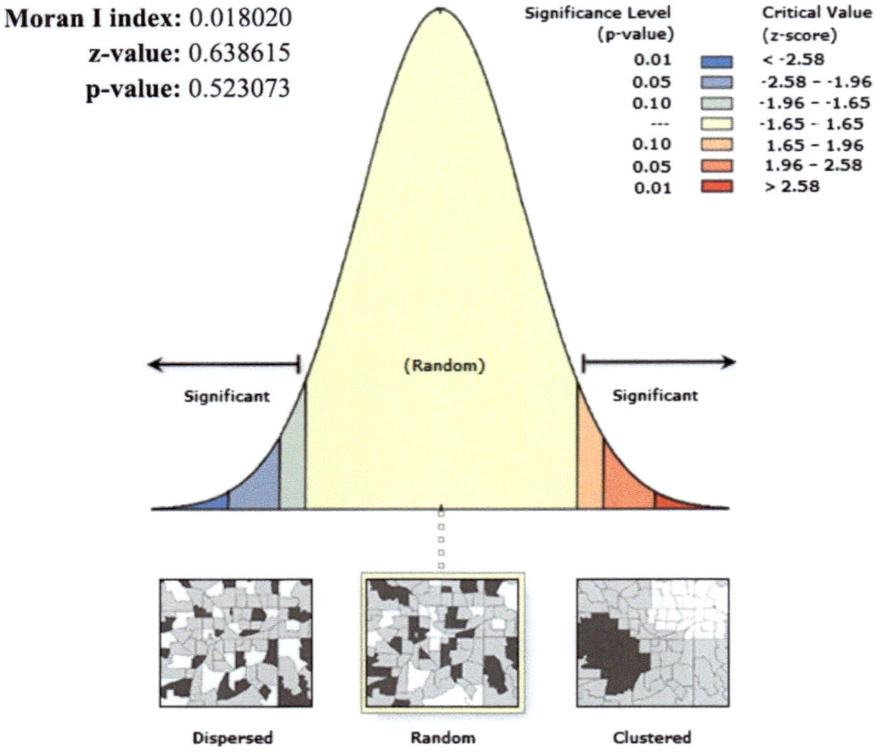

Fig. 4.8 Results of the spatial correlation analysis in 2019

synchronously and reduce execution disparities will enhance overall spatial autocorrelation and clustering possibilities.

1. Analysis of the Local Moran's *I*

Using STATA 18 software, the local Moran's scatterplots for urban safety resilience coupling coordination were created. As shown in Figs. 4.11, 4.12, 4.13, 4.14 and 4.15, the *x*-axis represents the standardized values of urban safety resilience coupling coordination for each year, and the *y*-axis represents their spatial lag values. The diagonal line indicates the regression line, with each number representing one of the 16 study areas: 1 = Chengdu, 2 = Zigong, 3 = Luzhou, 4 = Deyang, 5 = Mianyang, 6 = Suining, 7 = Neijiang, 8 = Leshan, 9 = Nanchong, 10 = Meishan, 11 = Yibin, 12 = Guang'an, 13 = Dazhou, 14 = Ya'an, 15 = Ziyang, and 16 = Chongqing. These visual representations aid in identifying local patterns and deviations, providing deeper insights into the dynamics of urban safety resilience across the economic circle.

Figures 4.11, 4.12, 4.13, 4.14 and 4.15 show that the local Moran's *I* are positive, demonstrating a spatial clustering characteristic and spatial association in the

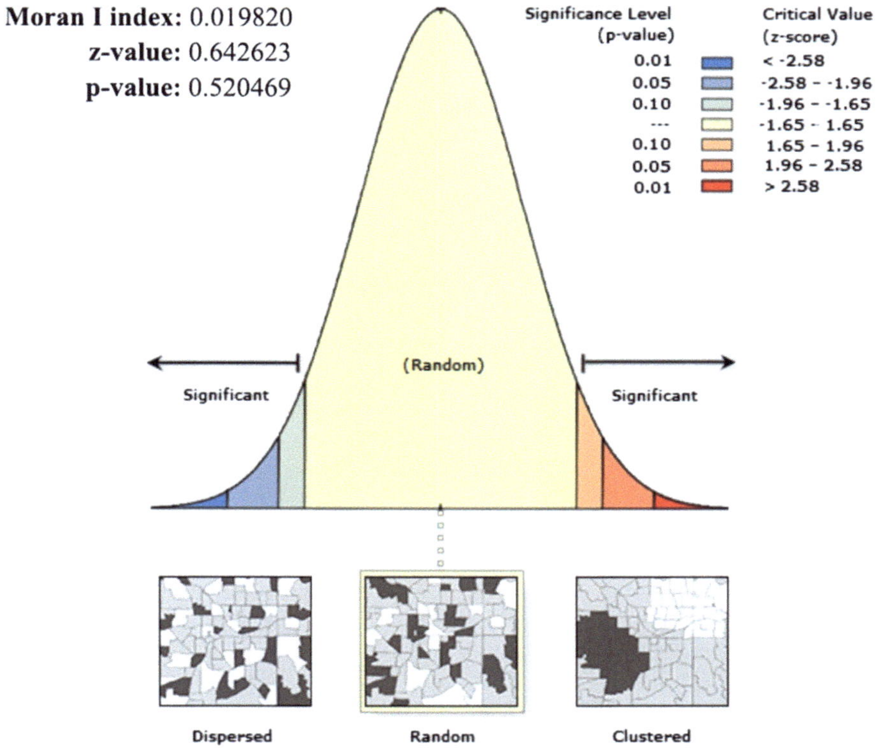

Fig. 4.9 Results of the spatial correlation analysis in 2020

coupling coordination development level of urban safety resilience. Provinces that are geographically next to each other have mutual influences and constraints in developing urban safety resilience coupling coordination. This emphasizes the complex spatial interaction between them. The local Moran's I indices initially increased and then decreased, with values shifting from -0.271 in 2017 to -0.177 in 2020, before dropping again to -0.213 in 2021. This pattern indicates that the spatial dependency and clustering of coupling coordination among geographically proximate cities strengthened initially and then weakened, with a slight decline over time.

The evolution of spatial distribution in urban safety resilience coupling coordination is driven not only by intrinsic factors but also significantly by the interactions and constraints of neighboring areas.

Figures 4.11, 4.12, 4.13, 4.14 and 4.15 shows the Moran scatterplots for coupling coordination. Table 4.1 illustrates each municipal area's spatial clustering characteristics and associated effects. Most cities generally display negative spatial autocorrelations in urban safety resilience coupling coordination. These cities are geographically located in the second and third quadrants, indicating high-low clustering phenomena.

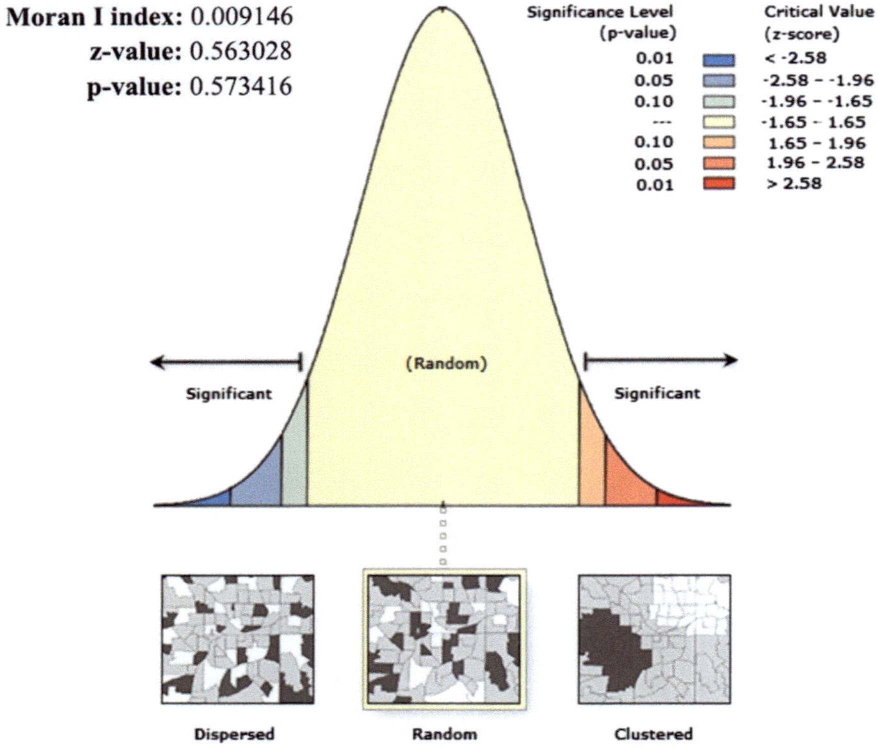

Fig. 4.10 Results of the spatial correlation analysis in 2021

The spatial autocorrelations of urban safety resilience coupling coordination in these municipal areas have evolved.

This evolving pattern suggests that while some regions demonstrate increasing resilience and improved coordination, others may still struggle or decline in their developmental progress. Addressing these disparities requires a detailed understanding of local conditions and broader regional dynamics. The information could inform targeted policy interventions and collaborative efforts to enhance overall urban safety and resilience across the economic circle.

The clustering types and their evolution in urban safety resilience coupling coordination can be examined in Moran's scatterplot. Here, four scenarios can be identified:

(1) High–High Clusters (H–H)

In the first quadrant, the central urban area and its neighbors have a high degree of urban safety resilience coupling coordination, showing a positive spatial correlation. From 2017 to 2021, a few cities showed the H–H clusters in the Chengdu-Chongqing Economic Circle, except Dazhou city in 2020. It indicates that most cities

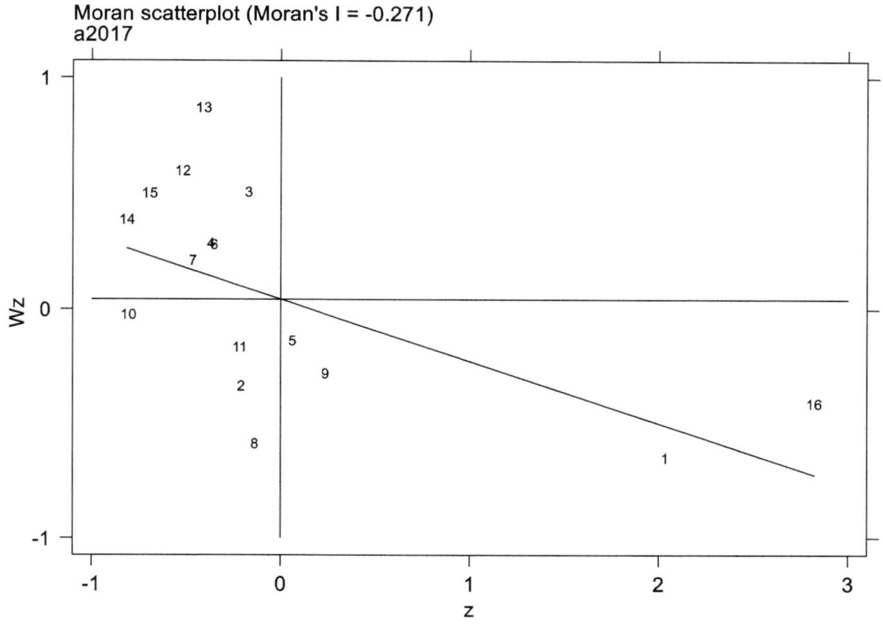

Fig. 4.11 Scatterplot of local spatial autocorrelation analysis in 2017

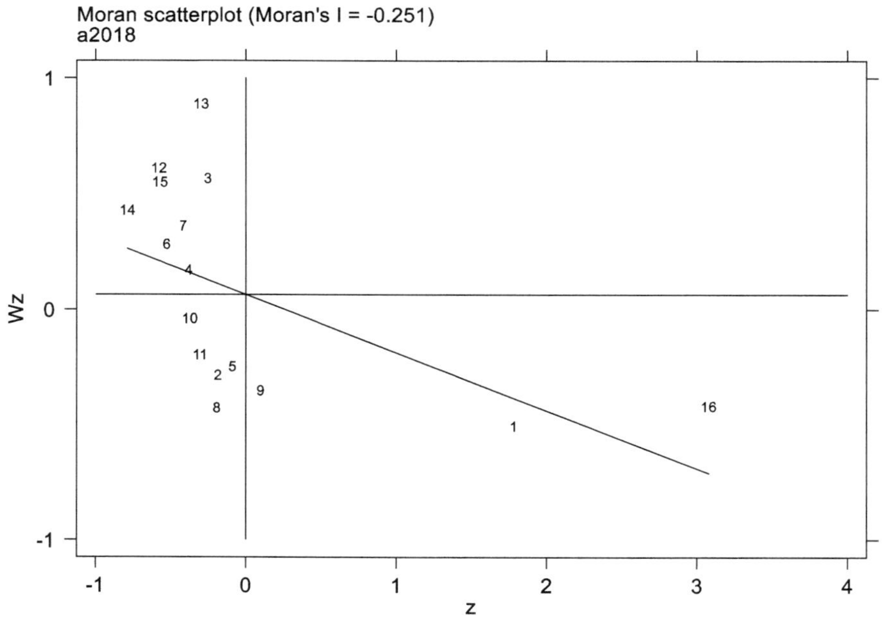

Fig. 4.12 Scatterplot of local spatial autocorrelation analysis in 2018

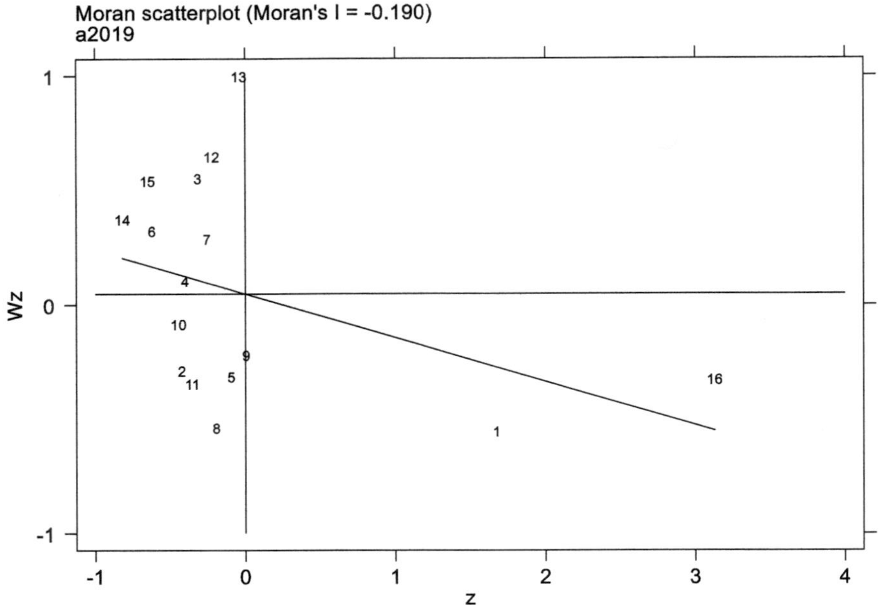

Fig. 4.13 Scatter plot of local spatial autocorrelation analysis in 2019

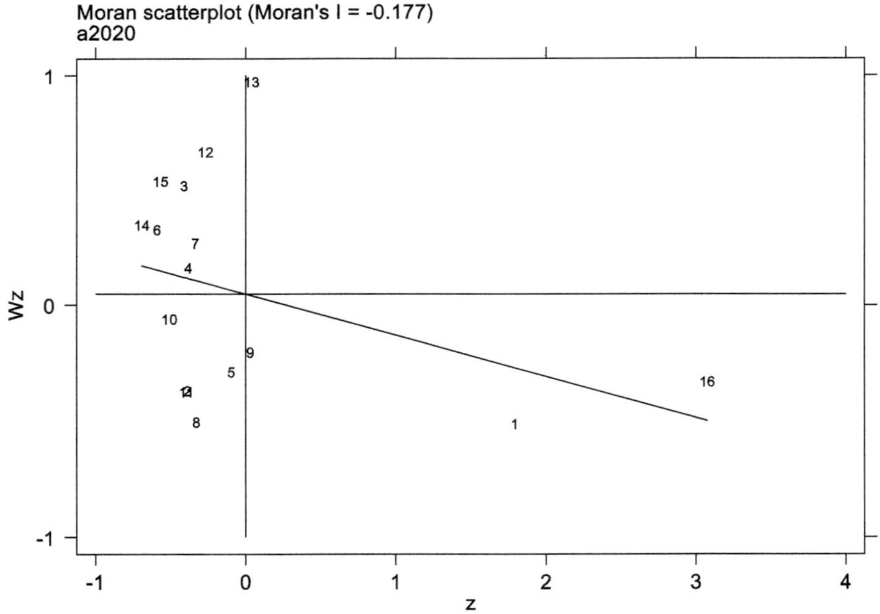

Fig. 4.14 Scatter plot of local spatial autocorrelation analysis in 2020

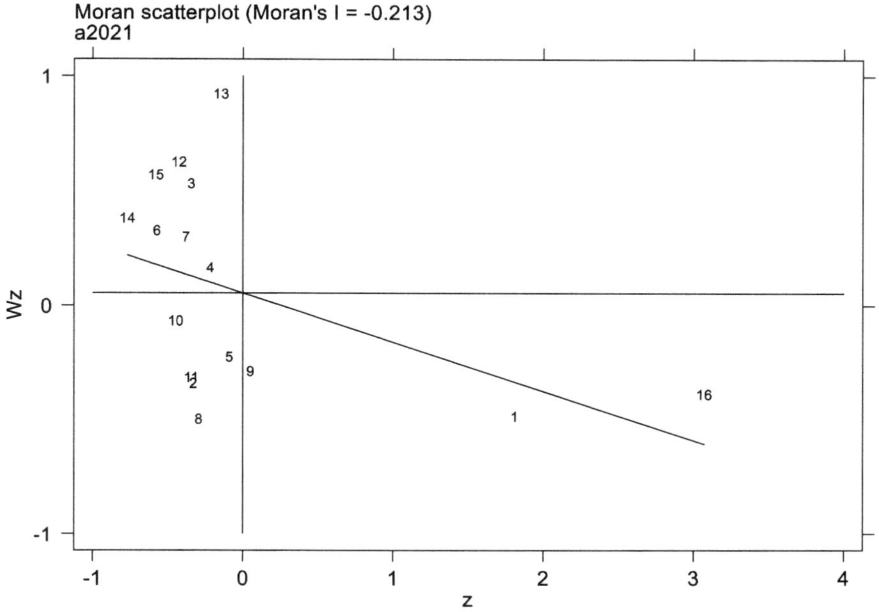

Fig. 4.15 Scatter plot of local spatial autocorrelation analysis in 2021

Table 4.1 Moran scatterplot corresponding to the coupling coordination degree of urban safety resilience

Quadrant	Agglomeration type	2017	2018	2019	2020	2021
I	H–H				Dazhou	
II	L–H	Luzhou, Deyang, Suining, Neijiang, Guang'an, Dazhou, Ya'an, Ziyang	Luzhou, Deyang, Suining, Neijiang, Guang'an, Dazhou, Ya'an, Ziyang	Luzhou, Deyang, Suining, Neijiang, Guang'an, Dazhou, Ya'an, Ziyang	Luzhou, Deyang, Suining, Neijiang, Guang'an, Ya'an, Ziyang	Luzhou, Deyang, Suining, Neijiang, Guang'an, Dazhou, Ya'an, Ziyang
III	L–L	Zigong, Leshan, Meishan, Yibin	Mianyang, Zigong, Leshan, Meishan, Yibin	Mianyang, Nanchong, Zigong, Leshan, Meishan, Yibin	Mianyang, Zigong, Leshan, Meishan, Yibin	Mianyang, Zigong, Leshan, Meishan, Yibin
IV	H–L	Chengdu, Mianyang, Nanchong, Chongqing	Chengdu, Nanchong, Chongqing	Chengdu, Chongqing	Chengdu, Nanchong, Chongqing	Chengdu, Nanchong, Chongqing

did not achieve high coupling coordination and positive spatial interactions during the observed period.

(2) Low–High Clusters (L–H)

In the second quadrant, the central urban area has a low coupling coordination degree, while its neighbors have a high coupling coordination degree, showing a negative spatial autocorrelation. The low-value areas are surrounded by higher-value neighboring units, forming a "depression" in safety resilience development. In 2017, Luzhou, Deyang, Suining, Neijiang, Guang'an, Dazhou, Ya'an, and Ziyang cities were classified as L–H areas, accounting for 50% of the total. In 2020, Dazhou transitioned from the L–H cluster to the H–H cluster and returned to being to the L–H cluster in 2021.

(3) Low–Low Clusters (L–L)

In the third quadrant, the central urban area and its neighbors have a low degree of urban safety resilience coupling coordination, showing a positive spatial correlation. In 2017, Zigong, Leshan, Meishan, and Yibin primarily showed the L–L clusters. In 2018, Mianyang shifted from the H–L cluster to the L–L cluster, increasing the number of such cities to five. In 2019, Nanchong became a member of this group, increasing the total number to six and accounting for 37.50% of the total number. In 2020, Nanchong moved from the L–L cluster to the H–L cluster and kept the classification in 2021.

(4) High–Low Clusters (H–L):

In the fourth quadrant, the central urban area has a high coupling coordination degree, while its neighbors have a low coupling coordination degree, showing a negative spatial autocorrelation. In 2017, Chengdu, Mianyang, Nanchong, and Chongqing showed the H–L clusters, accounting for 25% of the total. In 2018, Mianyang shifted from the H–L cluster to the L–L cluster, reducing the number of H–L cluster cities to three. In 2019, the change in Nanchong reduced the H–L cluster cities to two. In 2020, Nanchong returned to the H–L cluster, restoring the count to three cities. No other changes occurred in 2021.

From 2017 to 2021, cities within the H–H clusters of the Chengdu-Chongqing Economic Circle were sparse, with Dazhou being the only temporary example in 2020. Most cities did not achieve a high degree of urban safety resilience coupling coordination. At first, many cities were located in low–high clusters, suggesting that d even if these cities had lower degrees of coupling coordination, their surrounding cities performed better. Over time, this situation has changed as some cities experienced improvements or declines in safety resilience, reflecting the uneven development within the region. Furthermore, several cities and their neighbors continuously exhibited a low degree of safety resilience coupling coordination. The persistence and increase of this clustering phenomenon expose vulnerabilities in urban safety resilience construction in specific areas.

The imbalanced development among cities in terms of economy, society, and infrastructure, along with a highly centralized tendency of resource support, are the primary reasons for the uneven spatial distribution of urban safety resilience coupling coordination. Therefore, establishing a more effective regional coordination mechanism to promote the equitable distribution and sharing of resources, information, and policies within the Chengdu-Chongqing Economic Circle is crucial. Such measures will significantly enhance urban clustering effects, contributing to a more resilient and cohesive regional environment.

2. Clustering Map

Spatial clustering maps are instrumental in delineating regions with significant spatial aggregation. These can be categorized into five distinct types based on the clustering criteria: Not Significant, High–High, Low–Low, Low–High, and High–Low. This classification differs from the scatter plot, which primarily reflects the positions of all points across the four quadrants. The divisions in the scatter plot are based solely on the quadrant boundaries, with classifications made by observing whether a point lies within or is closest to a specific quadrant. Thus, the scatter plot analysis did not initially differentiate the "Not Significant" category (Anselin 1995).

Using ArcGIS 10.8, the LISA analysis was conducted from 2017 to 2021 to examine the coordination of urban safety resilience coupling within the Chengdu-Chongqing Economic Circle. The resulting LISA clustering maps are illustrated in Figs. 4.16, 4.17, 4.18, 4.19 and 4.20.

Fig. 4.16 Spatial clustering in 2017

Fig. 4.17 Spatial clustering in 2018

Fig. 4.18 Spatial clustering in 2019

Fig. 4.19 Spatial clustering in 2020

Fig. 4.20 Spatial clustering in 2021

These maps provide a nuanced visualization of how cities within the economic circle cluster in terms of urban safety resilience. The 'Not Significant' category, which was not evident in the scatter plots, includes areas where the spatial correlation is too weak to meet the statistical significance required for one of the other four categories. This addition helps clarify the spatial dynamics and connectivity or lack thereof, across the region, highlighting the areas where urban resilience efforts might be effectively concentrated or where further investigation is needed to understand underlying disparities.

As can be seen in Figs. 4.16, 4.17, 4.18, 4.19 and 4.20, Leshan City was the only one consistently identified as a "Low–Low" spatial cluster, indicating that both the city and its surrounding areas show a low degree of urban safety resilience coupling coordination. In 2018, Leshan's spatial clustering had become non-significant, a trend that extended across the Chengdu-Chongqing Economic Circle until 2020. During that year, Leshan and Neijiang City transitioned from non-significant to significant spatial clustering (i.e., Low–Low cluster), reflecting a persistent low degree of safety resilience coupling coordination in these cities and their surrounding regions. The shift in Neijiang from a significant Low–Low cluster to a non-significant may suggest some degree of change in the coupling coordination with its neighboring cities.

It can be inferred that Leshan and Neijiang City may have lower degrees of economic development than other cities, directly impacting their capacity to enhance urban safety resilience. Additionally, the regional cooperation mechanisms within the economic circle may not have effectively promoted synergistic development in cities like Leshan and Neijiang.

In conclusion, introducing investments and developing distinctive industries, strengthening the cities' economic development of cities can provide the economic foundation necessary for enhancing urban safety resilience. It is essential to increase the investment in infrastructure in these cities, particularly for public safety, transportation, and communication aspects. These measures can effectively boost the safety resilience of cities (e.g., Leshan and Neijiang), improving their spatial clustering characteristics and promoting balanced development within the Chengdu-Chongqing Economic Circle.

References

Anselin, L. 1995. Local indicators of spatial association—LISA. *Geographical Analysis* 27 (2): 93–115.

Rulian, W., L. Hongyi, and T. Fengj. 2021. The spatial distribution and its influencing factors of China 's national study travel. *Scientia Geographica Sinic* 41 (7): 1139–1148 (in Chinese).

Yingxia, P., G. Ying, M. Ronghua, et al. 2005. Analyzing regional economic disparities based on ESDA. *Geographical Research* 6: 965–974 (in Chinese).

Yong-jie, Z., and Q. Zhong-ying. 2011. Empirical research on spatial agglomeration of China's energy industry and the impact factors. In *2011 International Conference on Management Science & Engineering 18th Annual Conference Proceedings*, 978–987.
Zhigang, L., T. Shukun, L. Xiaoyan, et al. 2006. Research on the spatial distribution of Chinese innovative output: Spatial econometrics analysis based on province-level patent data. *Science of Science and Management of S&T* 8: 64–71 (in Chinese).

Chapter 5
Related Countermeasures and Suggestions

There are many cities similar to the Chengdu-Chongqing Economic Circle all over the world, such as the San Francisco Bay Area, Greater Tokyo Area, New York Metropolitan Area, Pearl River Delta, Jing-Jin-Ji (Beijing-Tianjin-Hebei), and the London Metropolitan Area, provide valuable comparative insights. Drawing on these parallels and the specific case study of the Chengdu-Chongqing Economic Circle, this book incorporates the findings from the evaluation of development levels and obstacles within urban safety resilience subsystems, coupling coordination measurements, and spatiotemporal evolution analysis. According to international frameworks such as the *Transforming our World: 2030 Agenda for Sustainable Development*, the *Paris Agreement*, and *the 100 Resilient Cities (100RC)*, this section proposes pathways and policy recommendations to promote the development of urban safety resilience and the coordinated development of its subsystems.

Optimization Paths and Countermeasures for Urban Safety Resilience Development

Building resilient cities was first proposed in the 1970s and gained widespread dissemination in 2010. It aimed to mitigate the ills of accelerated urbanization and the frequent natural disasters brought on by climate change. As time progressed, the concept of urban resilience began to incorporate socio-economic factors, community participation, and technological innovation. A subset of the broader urban resilience concept, urban safety resilience emphasizes the essential nature of safety in resilience planning. Safety is fundamental to all aspects of life, and enhancing urban safety resilience is crucial for boosting overall city resilience.

Y. Xiang et al., *Research on Coupling Coordination of Urban Safety Resilience Based on Disaster Prevention and Reduction*, Resilient and Friendly Living Environments,
https://doi.org/10.1007/978-981-96-1562-9_5

Considering multiple factors, the impact of urban scale and the city's disaster prevention and emergency management systems are vital. By addressing these aspects, cities can overcome obstacles in developing their safety resilience, fill gaps in their capabilities, and promote the healthy development of urban safety resilience. The comprehensive approach ensures that cities are better prepared to handle adversities and sustainably develop in harmony with their environments.

Promoting a "Human and Nature-Centric" Urban Development Strategy

The research indicates that safety perception and culture are significant impediments to urban personnel safety resilience, while construction projects hinder urban facility safety resilience. Essentially, the lack of a robust safety culture and extensive construction expansion without adequate safety measures hinder the development of urban safety resilience. The frequent occurrence of extreme natural disasters, whether due to human or natural factors, compromises sustainable urban development. Judy Bush suggested that urban planning from a purely human perspective failed to address many urban challenges comprehensively. Solutions incorporating non-human nature could promote social equity and address the trade-offs between time and space. For densely populated countries like China, considering urban development from a people and nature-centric perspective is more effective.

1. Enhance Basic Social Security to Alleviate the Urban Residents' Safety Crisis

Addressing the security needs of the urban unemployed can significantly enhance the general sense of safety among city dwellers. Despite the complexities of the global situation, which make it difficult to resolve unemployment fully, alleviating job pressure at the national level is crucial. Experience over the past fifty years in unemployment management internationally shows that public employment services (or employment counseling services) are the most cost-effective policy tool against unemployment. Establishing effective local labor market information platforms can significantly reduce frictional and structural unemployment. However, effective public employment services rely on reliable labor market information, necessitating increased government efforts in labor market research and coordination. Integrating apprenticeship systems with vocational rehabilitation mechanisms and combining employment subsidies with vocational training can maximally assist young and structurally unemployed individuals to integrate into the labor market. Employment assistance policies should balance welfare and economic rationality, digging into their developmental functions, such as promoting employment, not just focusing on welfare assistance. It is imperative to design policies that consider the feasibility of individuals obtaining employment through policy assistance while also considering the extent of help they need and their desire to participate in social production.

2. Address the Challenges Caused by an Aging Population

Sustaining social security systems is a long-term strategy to combat aging. Enhancing the welfare level accessible to the elderly over 65 years old, reducing multidimensional and health poverty rates, and maintaining the balance and stability of social security systems are essential. Appropriate financial policies should be adopted to alleviate the pressure on local financial sustainability during urban population transitions. For instance, local governments should adjust the indirect tax burden and pension and healthcare policies based on the rate of population aging, educational levels, and stages of economic development in different regions to mitigate risks to local finances during demographic transitions. Support the disease conditions and mental health of the elderly population, minimizing the negative impact of diseases on their life qualities. Promote early screening, treatment, and prevention of diseases among them and strengthen psychological counseling and care for this demographic, mainly focusing on empty-nesters and single or solitary elderly individuals. Encourage active participation in social activities to enhance their mental health. Enhance the role of family caregiving and proxy support in multi-dimensional poverty reduction.

3. Transforming the Construction Industry from Extensive Management to Enhanced Modernization

(1) Improving Social Security for Construction Workers and Fostering a High-quality Workforce

Building a resilient city requires sustainability goals, experienced managers, and skilled construction workers. Ensuring the labor rights and employment stability of construction workers can be achieved through a "National Construction Workers Information Integration Supervision and Service System (Platform)." Workers can receive worker qualifications for the construction industry through this platform if they pass a series of reviews, training, and assessments and meet the professional entry requirements. Establishing individual professional electronic records and providing dynamic tracking services can create a "protection circle" for these workers.

(2) Ensuring Occupational Safety and Health of Construction Workers

Guaranteeing construction workers' occupational safety and health is essential for attracting high-quality workers and cultivating a stable workforce. Government authorities should implement necessary measures to motivate enterprises to improve the working conditions of construction workers. Offering tax incentives and rewards to employers who meet the standards for construction site conditions throughout the project lifecycle, such as bonus points for quality project awards and upgrades in credit ratings, can significantly motivate employers to enhance their workers' occupational safety and health. Conversely, employers who fail to meet these standards should face penalties, with severe cases being placed on a "blacklist" and subject to cross-departmental joint punishment, such as market access restrictions or downgrading of qualifications.

(3) Addressing a Salary Delay Problem in Construction

Delays in the payment of wages to construction workers can cause significant delays in urban construction projects. To better guarantee the timely disbursement of salaries and benefits to construction workers, it is recommended that labor costs be separated from the construction payment guarantee and that a "Construction Workers' Salary Payment Guarantee" be established, managed, and guaranteed by banks.

(4) Promoting Industry Transformation and Upgradation through Digital Technology

Digital technology is central to the construction of smart cities, which has been shown to enhance urban resilience significantly. It is vital to focus on promoting digital technologies to build smart cities and enhance urban resilience, which can transform the construction industry. Continuously promoting the integration of smart cities with sustainable development, leveraging national strategic planning guidance, further perfecting market-oriented mechanisms for technological innovation, supporting industry upgrading and emerging industries, and improving the efficiency of resource allocation are essential. These efforts should aim to fully unleash the potential of human, financial, and material resources.

4. Enhancing Urban Forestry to Promote Coexistence between Humans and the Natural Environment

The development of urban forests can effectively reduce surface temperatures and enhance resistance to heatwaves, positively mitigating extreme heatwave disasters. By improving urban environmental quality, urban forestry initiatives promote better health among residents, decreasing healthcare expenses and lowering health costs. The ecological effects of urban forests form the foundation for other beneficial impacts, which can positively influence the urban ecological environment. During the construction and management of urban forests, government agencies should focus on guiding the comprehensive effects of these initiatives to promote the coordinated and sustainable development of the "natural-economic-social" aspects of cities.

It is crucial to be vigilant about the potential new imbalances caused by differences in ecological and environmental quality between regions and groups. Efforts should be made to avoid trapping economically underdeveloped areas and low-income groups in an environmental health poverty cycle. Governments need to fully recognize the varying impacts of environmental improvements on health expenditures across different regions and income groups. They should develop environmental and health service policies that favor and benefit underdeveloped areas and low-income populations to prevent them from falling into a multi-faceted poverty scenario encompassing environmental, health, and economic aspects, thereby exacerbating existing regional and group imbalances.

Addressing Developmental Shortcomings in Different Scale Cities Within the Chengdu-Chongqing Economic Circle

The Chengdu-Chongqing Economic Circle, driven by the "dual-core" cities of Chengdu and Chongqing, illustrates a significant disparity in economic development, infrastructure, and public services compared to other cities within the circle. These two megacities have far outpaced other regional cities, and the gap has not narrowed over time. This scenario is a classic example of the "Matthew Effect," where the wealthy continue to accumulate wealth, while the impoverished face difficulties in narrowing the gap. To counteract this and foster equitable development, a strategy focusing on the unique development of each city and promoting multi-city cooperation is crucial.

1. Enhancing Public Services to Improve Urban Population Protection Capabilities

The Chengdu-Chongqing Economic Circle displays a marked imbalance in developing urban personnel safety resilience, with Chengdu and Chongqing significantly outstripping other cities. This disparity can be attributed to the unique administrative resources available to provincial capitals and direct-controlled municipalities. However, the large resident populations of these megacities also give rise to numerous civil life issues. It is essential to increase investments in urban research, enhance the construction of modern urban agglomerations, improve the social security system, and ultimately enhance urban personnel safety resilience.

(1) Enhancing Urban Innovation and Research

As hubs for high-caliber talent, these cities must continue to use their influence to promote innovation-driven population aggregation. The government should increase investments in urban innovation, modify scientific research funding plans, and enhance support for top laboratories and critical research centers. By improving the financial systems for research and simplifying the fund verification processes for researchers, the initiative can boost the motivation of scientific personnel. The government should further support the construction of platforms for data equipment sharing and research outcome transformation, remove obstacles to knowledge sharing, optimize the allocation of scientific resources, reduce the costs of R&D, accelerate the transformation of research results, and promote effective linkage between production, education, and research. These approaches ensure the rapid development of local innovation clusters, which can enhance the city's capacity to attract and retain a population and provide a robust platform for attracting high-quality talent.

(2) Guaranteeing Industrial Aggregation's Positive Impact on Population Aggregation

Optimizing the urban traffic networks is essential to ensure that industrial aggregation positively impacts population aggregation. It can manage commuting needs effectively and guides the orderly relocation of traditional labor-intensive industries to suburbs or surrounding satellite cities, achieving a corporate operational model that separates headquarters from manufacturing sites. It is vital to promote the development of high-tech industries in urban centers, create a complementary urban functional model, and enhance specialization and collaboration within metropolitan areas. These measures can ensure the rapid development of small and medium-sized manufacturing enterprises.

(3) Accelerating the Integration of Social Security

As metropolitan infrastructure construction accelerates, government agencies should expedite the integration of social security measures to address the real issues brought by population concentration, such as off-site elderly care and medical treatment in different locations. This will fully leverage the contribution of non-local populations to local economic construction. The government should adopt diverse strategies to replenish social security funds and expand their scale while strengthening existing capital collection channels and increasing financial investment. Establishing dedicated fund management organizations and actively exploring a scientific asset management system for social security funds is also essential. Adhering to principles of prudence, ensuring safety, and adopting long-term, value, and responsible investment strategies to achieve asset preservation and appreciation are crucial. Moreover, it is necessary to strengthen fund managers' ideological and operational construction, expedite the formation of a professional asset management team with high levels of expertise and integrity, and complete the supervision mechanisms for the operation of social security funds. Establishing a mechanism for the public disclosure of social security fund information is vital to ensure the safety of the funds and prevent corruption and misappropriation.

Infrastructure construction and talent training are the most important for other small and medium-sized cities in the Chengdu-Chongqing Economic Circle. Strengthening the allocation of educational resources, improving the medical and health system, and strengthening basic community services should be adopted to improve the safety and resilience of urban residents.

(4) Strengthening Education Resource Allocation

Establishing a diversified financing system for higher education through legislative guidance, financial allocations, and policy standardization is critical. Encouraging universities to engage in external research collaborations and provide research services can generate economic returns and transform research outcomes.

(5) Building a Robust Primary Healthcare System

When Chinese higher-tier hospitals support lower-tier hospitals, they should focus on transferring high-quality medical resources downward while also training outstanding local medical talent. Given the varying disease spectrums across different regions, higher medical institutions should dispatch specialized doctors, general practitioners, and nursing teams according to the characteristics of diseases to enhance the level of medical services. They should introduce the latest standards for diagnosing and treating common, emergency, and referable diseases to grassroots medical institutions. Furthermore, higher-tier hospitals should organize relevant training based on general medical treatment, with the family doctor team playing a central role, establishing a collaborative mechanism to facilitate the joint involvement of leading hospitals and grassroots hospitals.

(6) Promoting Grassroots Community Democratic Governance

Implementing community consultation mechanisms that follow a six-step process: collecting opinions, determining issues, devising plans, announcing decisions, organizing consultations, and utilizing outcomes. Designating a specific day each month as "Community Consultation Day" to organize meetings with community residents can address local issues such as greenery maintenance and waste clearance through consultation, understand residents' living conditions, and then launch corresponding social service projects. Expanding channels for expressing community sentiments through platforms like WeChat public accounts and community social platforms can effectively stimulate widespread participation.

2. Strengthening Infrastructure to Enhance Urban Facility Security

In the Chengdu-Chongqing Economic Circle, the rate of infrastructure development significantly lags behind the economic growth rate, which poses challenges to the development of urban safety resilience. Increasing investment in infrastructure and boosting technological innovation is critical. Here are some targeted strategies:

(1) Enhancing Infrastructure Investment with a Focus on Efficiency

According to the locational advantages of different regions, optimizing the layout of infrastructure investments is essential. It involves addressing the deficiencies in infrastructure development and constructing a comprehensive support system. Investing in high-output efficiency smart city infrastructure, such as telecommunications and transportation facilities, is advisable. Such investments should fully capitalize on the role of infrastructure in driving economic growth.

(2) Advancing Intelligent Infrastructure Development

There should be an accelerated effort to develop intercity high-speed railways, urban rail transit, and highways. Coordinating cross-regional digital infrastructure connectivity advancement is crucial, breaking down market segmentation and monopolies and continuously enhancing the market integration of central cities and key urban clusters. Strong support measurements can upgrade modern information technologies like 5G and the Internet of Things (IoT). Promoting the evolution of information networks towards ubiquitous networks can provide intelligent capabilities to infrastructure, solidifying the interconnected foundations of various urban subsystems. These measurements would genuinely realize urban intelligence and significantly promote the development of urban facility safety resilience.

(3) Implementing Smart City Technologies

Smart city technologies improve efficiency and enhance the resilience and security of urban systems. Implementing advanced monitoring and automated management systems can help in real-time problem detection and resolution, which is crucial for maintaining the integrity and functionality of critical infrastructure.

(4) Encouraging Public–Private Partnerships (PPPs)

Public–private partnerships (PPPs) can be particularly effective to fund and manage these extensive infrastructure projects. These collaborations can bring in additional funding, innovative ideas, and efficient management practices from the private sector, which can help deliver construction projects on time and within budget.

(5) Fostering Community Involvement

Engaging the community in infrastructure planning and feedback processes ensures that the projects meet the residents' actual needs and helps garner public support for ongoing and future projects. Community involvement can also lead to more sustainable and user-friendly infrastructure development.

Enhancing Urban Multi-hazard Prevention and Emergency Response Capabilities

In the Chengdu-Chongqing Economic Circle, urban management safety resilience, particularly regarding material supply capabilities, has shown to be relatively weak. The fluctuating annual performance levels in various cities indicate the impact of recurring disasters such as floods, earthquakes, and droughts, significantly disrupting urban resilience development. Thus, tailoring disaster prevention and emergency management strategies to fit local conditions is critically important.

1. Establishing an Information-Based Emergency Response Platform

Creating an information-based emergency response platform, with the national government at the core to integrate both local and central levels, is a pivotal step towards enhancing disaster response capabilities. This strategic approach entails two main components:

(1) Optimizing Urban Multi-hazard Risk Management and Emergency Response Capacity Building

Utilizing the information-based emergency platform, consolidate data related to disaster risks, emergency materials, equipment, manpower, and organizational systems from various cities into a digitized format. This integration allows for the easy identification of regions where capabilities do not match risks, enabling timely measures, such as increasing investment in emergency capacity building. The platform facilitates real-time monitoring and management, ensuring that when a disaster strikes, all necessary information is readily accessible, thereby streamlining the response process.

(2) Establishing a Unified Assessment Standard for Disaster Emergency Capability

Develop a standardized mechanism for evaluating disaster response capabilities across cities. It involves setting uniform standards for cities to conduct and publish emergency capability evaluations regularly. These reports will drive cities to address any deficiencies in their emergency response infrastructure and ensure continuous tracking and assessment. Establishing these standards promotes consistency in measuring and improving emergency preparedness across regions and enhances accountability and effectiveness in emergency management (Table 5.1).

The construction of an information-based emergency platform not only helps comprehensively understand the emergency preparedness levels of various cities from a macro perspective but also aligns with the development trends of modernizing emergency management systems and capabilities. This platform is a critical tool in fostering a more resilient and responsive urban environment, prepared to efficiently manage and mitigate the impacts of multi-hazard events.

2. Strengthening Emergency Support Infrastructure

Emergency support is crucial for mobilizing human, material, financial, and technological resources to respond to sudden incidents. It includes the various resources needed at all stages of emergency management: mitigation, preparation, response, and recovery. Enhancing disaster emergency support infrastructure is vital for improving disaster emergency support capabilities.

Table 5.1 Benefits of an information-based emergency platform

Benefits	Specific description
Centralized resource management	Centralizes information about resources, reducing redundancies and ensuring that critical supplies are directed where they are most needed during emergencies
Enhanced communication	Improves communication between different levels of government and agencies involved in emergency response, ensuring that all parties have up-to-date information and can coordinate efforts more effectively
Data-driven decision-making	Provides a robust database that can be used for real-time decision-making and long-term planning, improving the overall strategic response to emergencies
Public information and alert system	It can be used to disseminate information to the public quickly and efficiently, ensuring that citizens are adequately informed and can take appropriate actions during disasters
Training and simulations	Facilitates continuous training programs and simulation exercises, using the collected data to create realistic scenarios that can help improve preparedness and response strategies

(1) Strengthening Urban Material Support

The capability to support emergency materials is a critical factor in improving the effectiveness of disaster emergency responses and meeting multi-hazard material supply needs. It also plays a crucial role in driving the operation of emergency management mechanisms and achieving emergency management goals. First, establishing and refining legislation related to emergency material reserves is essential to enhance emergency material support capabilities comprehensively. It includes setting clear regulations and bases for the procedures and rules of material reserves. Second, an assessment mechanism should be established to evaluate material support capabilities scientifically. Assessments of material support capabilities can identify potential risks in material reserves, allowing for timely adjustments to the reserve structure and improvements in reserve quality. It also facilitates the dynamic updating of reserved materials. Lastly, a demand-based reservation system must be established. By constructing earthquake disaster scenarios and assessing material needs, the types, quantities, and methods of material reserves can be determined to ensure that emergency material supplies meet the needs of earthquake disaster emergency response.

(2) Continuously Providing Urban Emergency Financial Support

Emergency financial support is a core component of the emergency management system, providing the support and foundation for the effective operation of disaster emergency management activities. To enhance financial support capabilities, increasing the budget for contingency funds and establishing an auditing and supervision mechanism is necessary. Contingency funds serve an irreplaceable protective role and are not merely a source of activity funds. Policies should encourage governments at all levels to appropriately increase the proportion of contingency funds extracted to ensure financial backups. Additionally, establishing an auditing and supervision mechanism is vital. This mechanism should strictly regulate the procedures for using contingency funds to avoid issues such as fund stagnation or misappropriation.

3. Optimizing an Emergency Response Plan System

Emergency response plans are the fundamental basis for taking action and decision-making during crises, and they provide a standard framework for ensuring legal assurance. By refining an emergency response plan system, precise guidance can be provided to address multi-faceted disaster risks effectively. First, enhancing the systematic nature, normative clarity, and operability of emergency plans is imperative. Building upon the national disaster response framework, all levels of government, communities, educational institutions, and businesses must escalate their commitment to human, material, and financial resources to draft these plans. These plans must comply with existing legal and regulatory standards and encompass provisions for communication, financial and material readiness, and rescue operations, thus ensuring the plans' thoroughness. Second, the execution of plan drills is crucial. These targeted training exercises test the implementation of command personnel roles and rescue teams' efficacy in responding to emergencies. Drills expose any deficiencies in the plans or emergency preparedness, prompting timely enhancements and continuous improvement in emergency handling capabilities. Departments must develop emergency drill schedules, design scenarios, and conduct related training. Drills should be executed per these plans and include internal and external evaluations, on-site and comprehensive debriefings to document the drills' processes, outcomes, and identified issues. Third, attention must be paid to the dynamic management of plans. A periodic revision mechanism for emergency plans should be established, and clear guidelines should be set for the timing, procedures, and conditions of plan updates. This approach ensures dynamic management of the plans, maintaining their vitality and effectiveness and prolonging their lifecycle.

By structuring a robust and adaptable emergency planning system, institutions can ensure rapid and competent responses to unforeseen disasters, significantly mitigating risks and enhancing community resilience.

Optimizing Pathways and Strategic Recommendations for Urban Safety Resilience Coupling Coordination Development

Strengthening the coupling and coordination of urban personnel, infrastructure, and management safety resilience is crucial in enhancing urban safety resilience and sustainable urban development. In general, there is significant potential for improvement in the urban safety resilience coupling and coordination within the Chengdu-Chongqing Economic Circle. The region exhibits noticeable developmental imbalances, a fact that has been verified using spatial autocorrelation analysis. The phenomenon of the "dual cores" of Chengdu and Chongqing dominating indicates that cities with better safety resilience coupling and coordination within the economic circle have not effectively influenced surrounding areas. Most cities have low urban safety resilience coupling and coordination degrees, which is not attributed to a single subsystem. Instead, it necessitates a complete enhancement in urban safety resilience to maximize their coupling and coordination grades.

For some cities, the low urban safety resilience coupling and coordination degrees are mainly due to the impact of urban personnel safety resilience. These cities could significantly influence their coupling and coordination degrees through practical measures to improve urban personnel safety resilience. Although their coupling and coordination degrees are higher for megacities, the factors affecting coordinated development vary, reflecting that megacities should tailor urban safety resilience coupling and coordination development measures according to local conditions.

Clarifying the Functions and Characteristics of Each City

Spatial evolution analysis reveals that cities' coupling and coordination trends within the Chengdu-Chongqing Economic Circle vary significantly. There is also a wide disparity in their coupling and coordination degrees. Identifying and enhancing the functional positioning of each city can significantly aid in promoting their coupling and coordination.

1. Optimize Urban Economic Development Directions to Consolidate the Regional Economic Base

Due to differences in resources, geographical locations, and historical backgrounds, cities have distinct economic development directions. The city's economic base determines various aspects of its urban system, making it crucial to strengthen the economic foundation to enhance city functions and characteristics. For central cities, there should be a focus on diversifying industries to drive and radiate economic benefits to surrounding cities. Opting for diverse industrial development strategies can create a comprehensive development advantage, helping to distribute development risks and construct a complex economic system.

Strengthening industrial specialization or a combination of specialization and diversification is a beneficial approach for smaller cities. For instance, cities with significant agricultural resources should promote the deep processing of agricultural products. Similarly, cities with abundant labor should develop industries that can capitalize on this workforce advantage. Smaller cities with relatively modest economic scales may experience industrial chaos and lower overall competitiveness if they aggressively pursue diversification. However, depending only on a single industry can make these cities more susceptible to economic shocks. Therefore, while focusing on industrial specialization, smaller cities should also consider diversifying their industries. Specialized and diverse economies are both essential and not mutually exclusive. The choice between them should be made based on the city type and stage of urban development.

By adopting these strategies, cities can more effectively define their roles within the regional economy, contributing to a more balanced and resilient urban network within the Chengdu-Chongqing Economic Circle. This approach not only leverages the unique strengths of each city but also ensures that economic growth is sustainable and inclusive, mitigating the risks associated with economic downturns and enhancing the overall resilience of the region.

2. Rational Planning of Social Service Facilities to Ensure Adequate Coverage

Under the control of aggregate indicators, the layout of social service facilities often adopts large-scale spatial patterns, which can lead to imbalances in spatial distribution and misalignment between ecological and social functions. It is essential to employ quantitative methods such as spatial analysis and mathematical analysis to assess urban social service facilities' spatial equity comprehensively. This analysis should be coupled with surveys and user satisfaction feedback to capture urban residents' service demands (Table 5.2).

By using these strategies, cities can overcome the challenges associated with the large-scale spatial layout of social services, achieving a more balanced and equitable distribution. This approach improves urban residents' quality of life and enhances urban environmental functionality and sustainability.

3. Developing a Green Circular Economy to Enhance Urban Environmental Resilience

The global trend in urban development is increasingly focusing on the green circular economy, a model that promotes sustainability by minimizing waste and encouraging the reuse of resources. Enhancing urban environmental resilience can be significantly advanced by promoting green lifestyles among residents, such as green living, green consumption, and green mobility (Table 5.3).

By adopting these strategies, cities can make significant strides in developing a green circular economy that conserves resources and enhances urban environmental resilience. This approach ensures that cities are better prepared to handle environmental challenges and can provide their residents with a healthier, more sustainable living environment.

Table 5.2 Strategies for optimizing spatial layout of social services

Strategies for optimizing spatial layout of social services	Details
Utilize data-driven approaches	Conduct thorough spatial and mathematical analyses to identify areas where social service facilities are lacking or disproportionately distributed. The data-driven approach helps understand the current disparities and plan interventions more effectively
Engage with the community	Gather input from residents through surveys and feedback mechanisms to understand their needs and satisfaction levels with existing facilities. This bottom-up approach ensures that the planning and development of social services are closely aligned with the actual requirements of the community
Implement regulatory adjustments through urban planning	Use the insights gained from data analysis and community feedback to make informed decisions in urban planning. Adjust the spatial layout of social services to optimize their distribution across the city, aiming for equitable access for all residents
Emphasize an "upward adjustment" planning philosophy	Adopt an adaptive planning model that allows for ongoing adjustments based on evolving needs and feedback. The model supports the dynamic nature of urban development and helps balance urban spaces' ecological and social functions
Ensure equitable provision of social services	Strive for fairness in the provision of social services by ensuring that all areas of the city, particularly underserved and disadvantaged neighborhoods, have adequate access to these essential services

Table 5.3 Strategies for fostering a green circular economy

Strategies for fostering a green circular economy	Details
Establish "green mobility" demonstration cities	Set up demonstration cities to lead by example in green mobility. These cities would serve as models for promoting low-carbon and eco-friendly travel choices among their residents and in non-demonstration cities. Strengthen the construction and policy implementation in these cities to ensure comprehensive adoption of green travel initiatives
Enhance public awareness of green practices	Intensify efforts to educate urban residents about environmental protection and the importance of green practices. It includes increasing awareness of green mobility, energy conservation, and emission reduction. Aggressive promotion of healthy lifestyle choices can foster the adoption of green habits among city dwellers
Promote sustainable consumption	Encourage residents to engage in green consumption by providing incentives for purchasing eco-friendly products and implementing policies supporting sustainable daily life practices. It could include tax rebates, subsidies, or recognition programs for individuals and businesses demonstrating commitment to sustainable practices
Guide green consumer behavior	Encourage residents to engage in green consumption by providing incentives for purchasing eco-friendly products and by implementing policies that support sustainable practices in daily life. This could include tax rebates, subsidies, or recognition programs for individuals and businesses that demonstrate commitment to sustainable practices
Implement supportive infrastructure	Develop infrastructure that supports green living, such as comprehensive recycling systems, widespread availability of green spaces, and enhanced accessibility to public transportation. These infrastructures make green choices convenient and appeal to the general populace

Technological Innovation Leading Safety Resilience

1. Utilizing Digital Technology to Support Urban Emergency Response

Information is no longer secluded or suppressed in the era of big data and information. New media and self-media platforms, such as smartphones and live streaming, facilitate rapid information flow. The failure to capture incidental information promptly can transform it into a crisis, producing severe consequences. Focusing on collecting public information is crucial since the public's channels are often more sensitive and perceptive than governmental ones, leading to potentially delayed government responses that may not meet public expectations (Table 5.4).

Table 5.4 Strategies for enhancing emergency response with technology

Strategies for enhancing emergency response with technology	Details
Integration of public safety resources	Utilize existing data and public information systems to enhance the efficiency of emergency alert services. It can significantly improve the timeliness and consistency of alert dissemination, reducing the economic losses and casualties caused by disasters
Establishment of a unified emergency database	All levels of government and departments should appoint personnel dedicated to emergency data management. Implementing unified collection standards will facilitate the gathering, analyzing, and utilizing of emergency data. Regular integration, maintenance, and data updates are essential for efficient emergency management
Emphasis on historical case analysis	Gathering and analyzing historical cases of similar incidents can guide government agencies in scientifically managing ongoing situations. For instance, meteorological departments often compare the path, strength, and direction of incoming typhoons with historical typhoons to analyze potential disaster impacts and defensive measures
Improvement of governmental early warning systems	Before disaster strikes, governments must issue timely, effective, and accurate, credible, authoritative early warnings. These warnings should facilitate disaster risk prevention and control, guide the public to take emergency precautions, and minimize damage, enhancing service provision

By integrating digital technology into emergency response frameworks, governments can harness the power of data to anticipate and mitigate the impacts of disasters more effectively. This approach safeguards public safety and ensures that the community remains resilient in emergencies.

2. Proactive Prevention: Establishing Urban Safety Monitoring and Alert Systems

Crises often herald their approach with significant precursor signals. Gathering and analyzing these signals can preempt sudden incidents. Accidents stem from an accumulation of minor oversights, which could escalate into a chain reaction. Although a single minor issue might not lead to an accident, a series could precipitate a severe crisis or directly cause an accident. A crucial component of emergency management involves leveraging big data to gather and categorize crisis information. This allows for identifying, managing, and mitigating potential hazards or crises at their nascent stages.

However, establishing a risk information advisory system is insufficient to prevent crises; developing an early warning system is also imperative. Once all safety measures are in place, real-time alerts can significantly mitigate the abruptness of crises, providing governments and the public sufficient time to prepare defenses and substantially reduce potential loss of life and property. Risk information advisory and crisis alert mechanisms are critical to enhancing emergency management capabilities. During the global COVID-19 pandemic, local governments required extensive supplies, including medications, medical equipment, transportation, and medical experts. Countries without a robust crisis advisory and alert system found their cities plunged into chaos when the crisis struck. An effective warning system allows for the orderly implementation of emergency measures in cities under crisis, efficiently handling various emergencies.

Strengthening Urban Infrastructure for DRR and Planning

1. Reducing Post-disaster Loss Through Comprehensive Disaster-Adaptive Measures

Urban infrastructure resilience reflects a city's capacity to respond to disasters. Compared to traditional DRR strategies that focus on reducing post-disaster losses, modern approaches promote a holistic, disaster-adaptive process throughout all stages of disaster management. China's central government has placed high importance on DRR and has made significant strategic decisions since the 18th National Congress. The country's new DRR philosophy, which includes "two persistences and three shifts," emphasizes prevention, the integration of prevention, resistance, and rescue efforts. It also emphasizes a shift from post-disaster relief to pre-disaster preparedness, from dealing with single disaster types to comprehensive risk reduction, and from merely reducing disaster losses to mitigating disaster risks.

Chinese urban DRR efforts have evolved from focusing on post-disaster relief to reducing losses post-disaster and improving urban resilience under comprehensive disaster-adaptive measures. The philosophy of "comprehensive disaster-adaptive DRR" stresses that each phase of the disaster lifecycle, including disaster prevention, response during the disaster, and post-disaster recovery, to be able to adapt to disasters effectively.

The pre-disaster phase in urban infrastructure can be strengthened through mechanism enhancement, creating model communities, education, and disaster risk analysis and assessment. Disaster risk analysis and assessment typically involve evaluating the causative factors of disasters, analyzing exposure levels, and assessing vulnerabilities. Regarding disaster prevention mechanisms, it is crucial to follow the principles of "prevention is primary, better safe than sorry," continuously strengthen leadership, assign clear responsibilities, and perfect emergency disaster plans. At the grassroots level, efforts should include building DRR model areas, enhancing DRR education and training, and raising public awareness and self-rescue capabilities.

Comprehensive adaptation relies on holistic collaboration in the in-disaster response process. A holistic, collaborative mechanism organizes the strengths of various levels and departments, promotes societal involvement, and combines the functions of different sectors to form a governance and collaborative work pattern. This modern approach to urban DRR is crucial as it involves the active involvement of society and government sectors in responding to disasters collectively. The strategy improves emergency response capabilities and is critical to urban infrastructure response.

The post-disaster recovery phase is a vital safeguard for comprehensive disaster-adaptive urban infrastructure. This phase involves returning the city to normalcy, learning from experiences, and improving the city's DRR capabilities at the infrastructure level.

Therefore, under the resilience concept, the entire process of urban infrastructure disaster prevention and reduction focuses on comprehensive disaster management and considers enhancing urban infrastructure DRR capabilities as integral to the approach.

2. Increase Funding for Infrastructure Maintenance

To effectively maintain the established DRR infrastructure, governments must prioritize and increase the allocation of resources and manpower dedicated to maintenance activities. Governments can promote a robust investment environment by establishing incentive mechanisms primarily driven by socio-economic strategies, which attract social capital and diversify investment sources, facilitating a comprehensive advancement of the DRR system.

During infrastructure construction and maintenance, innovative investment models should be aligned with market orientations. By integrating favorable national policies and well-established financing models, there should be a strategic plan to attract social capital involvement in infrastructure construction and maintenance, enhancing the maintenance and management level. The Build-Operate-Transfer (BOT) of the PPP financing model can secure additional funding for infrastructure management. This broadens the range of financial sources available for maintenance activities and lays a solid foundation for the ongoing upkeep of critical infrastructure.

3. Implement Information-Driven Maintenance Management

Once infrastructure is established, it enters a phase of long-term maintenance management. As modernization advances, digital and information technologies have become increasingly widespread. It is crucial to integrate information and digital technologies to develop a comprehensive urban infrastructure management database and platform to enhance the technical level of urban infrastructure management and maintenance. Establishing an infrastructure management system focused on early warning and maintenance response, enabling timely detection and resolving emerging issues, thus reducing disaster-related damages.

Adopting information-driven management approaches, paired with modern integrated diagnostic equipment, allows for the continuous monitoring and assessment of infrastructure. This proactive surveillance helps in the early identification of potential issues, facilitating timely repairs and reducing the likelihood of structural damages during the operational lifecycle of the infrastructure. This strategy extends the durability of the facilities and ensures their optimal functionality in disaster-prone scenarios, reinforcing the city's resilience against potential calamities.

Conclusion

As a specific manifestation of urban resilience in the security domain, urban safety resilience plays a crucial role in disaster prevention and mitigation. The increasing involvement of scholars in studying urban safety resilience highlights the significance of multidisciplinary and multi-sectoral collaboration due to its implications. Urban safety resilience comprises complex subsystems, and the interwoven relationships between their subsystems make it challenging to find a universally beneficial solution. This paper adopts the coupling coordination models and spatial analysis techniques from a systemic coupling perspective. A case study has been conducted on the temporal and spatial evolution of urban safety resilience coupling coordination in the Chengdu-Chongqing Economic Circle from 2017 to 2021. The results offer region-specific recommendations for enhancing urban safety resilience, providing strategic insights to advance sustainable development and establish a more sustainable and livable urban environment. The main conclusions are as follows:

1. This book has developed a comprehensive index system for evaluating urban safety resilience coupling coordination, including personnel safety, facility safety, and management safety dimensions. Based on panel data from 16 cities in the Chengdu-Chongqing Economic Circle from 2017 to 2021, an empirical test of urban safety resilience coupling coordination was conducted. The index system weights were calculated using an improved entropy method, highlighting the number of people covered by unemployment insurance, the number of people covered by work-related injury insurance, and the level of basic medical insurance for urban workers that contributed the most to urban personnel safety resilience. The level of urban traffic lighting facilities, urban drainage facilities, and the reserve level of construction industry personnel contributed the most to urban facility safety resilience. The comprehensive production capacity of the water supply, the financial support level of transportation, and the supply level of emergency materials were identified as making the most significant contributions to urban management safety resilience.

Y. Xiang et al., *Research on Coupling Coordination of Urban Safety Resilience Based on Disaster Prevention and Reduction*, Resilient and Friendly Living Environments,
https://doi.org/10.1007/978-981-96-1562-9

2. A definite interactive coupling relationship exists among the various subsystems of urban safety resilience. Overall, the development trends of urban safety resilience subsystems in the Chengdu-Chongqing Economic Circle are stable, with only minor variations observed in some cities. Specifically: (1) Urban Personnel Safety Resilience: During the research period, the development index for urban personnel safety resilience in the Chengdu-Chongqing Economic Circle showed a stable upward trend. The average value of the index rose from 0.185 at the beginning of the period to 0.228 by the end. Chengdu, Chongqing, Nanchong, Dazhou, and Mianyang city ranked in the top five for personnel safety resilience, while Yibin, Ya'an, Suining, Guang'an, and Ziyang city ranked at the bottom with less optimistic growth rates. (2) Urban Facility Safety Resilience: The development index for urban facility safety resilience grew more rapidly during the study period, rising from 0.223 to 0.294. Chongqing, Chengdu, Nanchong, Suining, and Mianyang city were the top-ranked cities, while Deyang, Dazhou, Meishan, Neijiang, and Ziyang city exhibited higher annual growth rates with lower ranking. (3) Urban Management Safety Resilience: The development index for urban management safety resilience also showed an upward trend during the research period, increasing from 0.229 to 0.260. Chongqing, Chengdu, Nanchong, Mianyang, and Yibin city were at the forefront, while Zigong, Suining, Ziyang, Meishan, and Ya'ancity were at the bottom.

   The development of the various urban safety resilience subsystems across the 16 cities of the Chengdu-Chongqing Economic Circle displays spatial consistency, primarily demonstrating a "dual-core dominance" pattern. This pattern highlights the central role of Chengdu and Chongqing as critical hubs of resilience, influencing surrounding areas and leading the overall resilience efforts within the circle. The data highlights the importance of enhancing urban resilience through targeted improvements in personnel, facilities, and management safety across all cities, especially those lagging in these areas. This comprehensive approach will address the disparities and strengthen the collective resilience of the entire economic circle.
3. Analysis of the obstacles within each subsystem of urban safety resilience reveals specific challenges that impede the progress of resilience in cities: (1) Urban Personnel Safety Resilience: The most significant obstacle is the sense of safety and safety culture. It implies that the perception and cultural framework surrounding safety in cities are critical factors that need enhancement to improve the overall resilience of city inhabitants. (2) Urban Facility Safety Resilience: The primary obstacle is related to construction engineering. Issues in building and infrastructure projects can significantly impair the ability of urban systems to withstand and recover from adverse events, indicating a need for stricter standards, better quality control, and innovative construction practices. (3) Urban Management Safety Resilience: The most significant obstacle is the material supply level. It refers to providing necessary resources during emergencies, crucial for effective disaster response and recovery.

The impact of these obstacles on urban safety resilience systems across different regional cities does not show significant variance, and these impacts have not changed considerably over time. This consistency indicates that although the Chengdu-Chongqing Economic Circle has recognized these issues, the measures implemented to address these vulnerabilities in urban safety resilience still require strengthening.

4. There is a need for improvement in the overall development of coupling coordination degree. The coupling coordination measurement results indicate that although the overall development condition of the urban safety resilience subsystems within the Chengdu-Chongqing Economic Circle is satisfactory, the coupling coordination development faces certain challenges. The overall degree of urban safety resilience coupling coordination increased from 0.399 in 2017 to 0.436 in 2021, and the classification improved from mild maladjustment recession to approaching maladjustment recession. This period can be divided into the "Chengdu-Chongqing Urban Agglomeration" phase (2017–2019) and the "Chengdu-Chongqing Dual-city Economic Circle" phase (2020–2021). However, the current development of coupling coordination degree remains relatively low. Specifically, as China's demographic dividend continues to decline, there appears to be a narrowing gap between urban personnel safety resilience, urban facility safety resilience, and urban management safety resilience. This change is minimal when considering the long-term development and could even be described as relatively static. Improvements in facility resilience and management resilience are the primary drivers for enhancing the degree of coupling coordination. Further increasing the growth rates of these two factors will significantly enhance the coupling coordination degree.
5. The coupling coordination across different regions shows signs of being insufficient and uneven. On one hand, the development of coupling coordination within the Chengdu-Chongqing Economic Circle is inadequate. Apart from Chongqing and Chengdu, the coupling coordination degrees of other cities range from 0.297 to 0.443, placing them between moderate maladjustment recession and approaching maladjustment recession. On the other hand, the development of coupling coordination within the circle is uneven. Chongqing, Chengdu, Nanchong, and Mianyang are the "first tiers" of urban safety resilience coordination development. Dazhou, Leshan, Luzhou, Zigong, Yibin, Deyang, Neijiang, and Guang'an city with average coupling coordination scores ranging from 0.36 to 0.4, form the "second tier." Suining, Ziyang, and Ya'an city are in the "third tier." The variation in coupling coordination among these cities is significant and reflects the "dual-core dominance" pattern also observed in the development levels of various subsystems.

   Moreover, the kernel density estimation analysis from 2017 to 2021 indicates a distinct division in the coordination of urban safety resilience coupling. There is no apparent tendency towards reducing these disparities. This polarization highlights the necessity for targeted developmental strategies and interventions customized to each city tier's specific needs and conditions.

6. The spatial distribution shows that the "dual-core dominance" of the Chengdu-Chongqing Economic Circle's urban safety resilience coupling coordination exhibits extremely distinct spatial differentiation characteristics. The temporal evolution across the regions shows a fluctuating trend without forming significant spatial clustering. Specifically, the Moran's I and *z*-values for the 16 regions studied are not ideal. Despite a clear upward trend, it is evident that there has not been significant spatial clustering of urban safety resilience coupling coordination within the Chengdu-Chongqing Economic Circle.

   The Moran scatterplots predominantly show the L–H and L–L clustering characteristics, indicating strong spatial associations. However, these plots also reflect that most cities do not exhibit an increasing trend in the H–H and L clustering due to low coupling coordination degrees. Additionally, the LISA clustering maps provide evidence that there is no significant spatial clustering of urban safety resilience coupling coordination. It emphasizes the importance of enhancing coupling coordination degrees and promoting synergistic development among cities.

Appendix A
Urban Safety Resilience Coupling Coordination Degree Evaluation Indicator System Survey

Dear Expert:

Hello! Our research team is currently working on the project "Urban Safety Resilience." The purpose of this questionnaire is to construct an evaluation indicator system for urban safety resilience in terms of coupling and coordination. Please mark "√" next to the indicators you consider appropriate and answer the related questions. Completing this questionnaire will take approximately 5 min of your valuable time, and we deeply appreciate your participation.

1. Your Basic Information

Employer: Administrative Department () University/Research Institute () Company ()

Job Position: Title:

2. Selection of Indicators for the Evaluation of Urban Safety Resilience Coupling Coordination Degree

The evaluation indicator system for urban safety resilience coupling coordination degree includes both quantitative and qualitative indicators. Please mark "√" next to the factors you consider appropriate. Detailed content can be seen in the table below.

Serial number	Indicator layer	Whether it is appropriate and adjustments are made
1	The density of permanent residents in the built-up area	
2	The level of basic medical insurance for urban workers	

(continued)

Y. Xiang et al., *Research on Coupling Coordination of Urban Safety Resilience Based on Disaster Prevention and Reduction*, Resilient and Friendly Living Environments,
https://doi.org/10.1007/978-981-96-1562-9

(continued)

Serial number	Indicator layer	Whether it is appropriate and adjustments are made
3	The proportion of temporary residents	
4	The proportion of disabled population	
5	The level of reserve health technical personnel in the city	
6	The reserve level of urban health institutions	
7	The number of police officers in ten thousand	
8	The number of firefighters in ten thousand	
9	The level of social organization units	
10	The number of people covered by unemployment insurance	
11	The number of people covered by urban old-age insurance	
12	The number of people covered by work-related injury insurance	
13	Commercial insurance density	
14	The per capita emergency deposit level	
15	Urbanization rate	
16	Proportion of land area in safety-vulnerable areas	
17	Proportion of earthquake-resistant buildings	
18	The level of personnel reserves in the construction industry	
19	Per capita urban road area	
20	The level of urban traffic lighting facilities	
21	The penetration rate of wired broadcasting and television	
22	The level of urban drainage facilities	
23	The per capita refuge area	
24	Public coverage of meteorological disaster monitoring, forecasting, and warning information	
25	Building area of disaster relief reserve institution warehouses per ten thousand people	
26	The number of beds in medical and health institutions per ten thousand people	
27	The green coverage rate	
28	The death rate per million people due to disasters	
29	The proportion of annual direct economic losses due to disasters to the region's gross domestic product	
30	The level of financial support for public safety	
31	The level of financial support for medical and health care	

(continued)

(continued)

Serial number	Indicator layer	Whether it is appropriate and adjustments are made
32	The level of financial support for transportation	
33	The comprehensive production capacity of water supply	
34	The level of grain supply	
35	The level of emergency material supply	

Appendix B
Second Round of Indicator Selection Expert Survey Form

Dear Expert:

Hello! Our research team is currently conducting the "Urban Safety Resilience" project. The purpose of this questionnaire is to analyze the importance of urban safety resilience coupling coordination evaluation indicators to the development of urban safety resilience. Please mark "√" under the appropriate importance score options based on the indicator system table and the indicator description table. The scores range from 1 to 10, with the importance of the indicators increasing accordingly. Completing this questionnaire will take approximately 10 min of your valuable time, and we deeply appreciate your participation.

1. Your Basic Information

Employer: Administrative Department () University/Research Institute () Company ()

Job Position: Title:

2. Selection of Indicators for the Evaluation of Urban Safety Resilience Coupling Coordination Degree

The specific content of the urban safety resilience coupling coordination evaluation indicators can be seen in the table below. Please mark "√" next to the importance score you deem appropriate.

Y. Xiang et al., *Research on Coupling Coordination of Urban Safety Resilience Based on Disaster Prevention and Reduction*, Resilient and Friendly Living Environments,
https://doi.org/10.1007/978-981-96-1562-9

Indicator layer	Importance score									
The density of permanent residents in the built-up area	1	2	3	4	5	6	7	8	9	10
The level of basic medical insurance for urban workers										
The proportion of temporary residents										
The number of newly added higher education talents										
The level of reserve health technical personnel in the city										
The reserve level of urban health institutions										
The level of social organization units										
The number of people covered by unemployment insurance										
The number of people covered by urban old-age insurance										
The number of people covered by work-related injury insurance										
The per capita emergency deposit level										
Urbanization rate										
Proportion of land area in safety-vulnerable areas										
The level of personnel reserves in the construction industry										
Per capita urban road area										
The level of urban traffic lighting facilities										
The penetration rate of wired broadcasting and television										
The level of urban drainage facilities										
Per capita refuge area										
Building area of disaster relief reserve institution warehouses per ten thousand people										
The number of beds in medical and health institutions per ten thousand people										
The green coverage rate										
The death rate per million people due to disasters										

(continued)

(continued)

Indicator layer	Importance score								
The density of permanent residents in the built-up area	1	2	3	4	5	6	7	8	9	10
The proportion of annual direct economic losses due to disasters to the region's gross domestic product										
The level of financial support for public safety										
The level of financial support for medical and health care										
The level of financial support for transportation										
The comprehensive production capacity of water supply										
The level of grain supply										
The level of emergency material supply										

Appendix C
Survey on the Relative Importance of Urban Safety Resilience Coupling Coordination Evaluation Indicators

Dear Expert:

Hello! Our research team is currently conducting the "Urban Safety Resilience" project. The purpose of this questionnaire is to determine the weights of the urban safety resilience coupling coordination evaluation indicator system. Please mark "√" next to the indicators you consider appropriate and answer related questions. Completing this questionnaire will take approximately 5 min of your valuable time, and we deeply appreciate your participation.

1. Your Basic Information

Employer: Administrative Department () University/Research Institute () Company ()

Job Position: Title:

2. Determination of Weights for Urban Safety Resilience Coupling Coordination Evaluation Indicators

The table below involves the urban safety resilience coupling coordination evaluation indicator system. On either side of the following tables are the evaluation indicators, with the middle part of the questionnaire reflecting the relative importance scores between these indicators. Please mark "√" under the appropriate scoring result based on the relative importance of the indicators on either side. The table shows the relative importance of indicator X to Y. This survey uses a 5-point scale, where scores from 1 to 5 indicate increasing importance of indicator X to Y, and scores from 1 to 1/5 indicate decreasing importance of indicator Y to X.

Y. Xiang et al., *Research on Coupling Coordination of Urban Safety Resilience Based on Disaster Prevention and Reduction*, Resilient and Friendly Living Environments,
https://doi.org/10.1007/978-981-96-1562-9

What is the relative importance of each of the following sets of pairwise comparisons of elements to "urban personnel safety resilience?"

X	Importance comparison									Y
	1/5	1/4	1/3	1/2	1	2	3	4	5	
Basic population attributes										Social participation readiness
Basic population attributes										Safety perception and safety culture
Social participation readiness										Safety perception and safety culture

What is the relative importance of each of the following sets of pairwise comparisons of elements to "urban facility safety resilience?"

X	Importance comparison									Y
	1/5	1/4	1/3	1/2	1	2	3	4	5	
Construction engineering										Transportation facilities
Construction engineering										Lifeline engineering facilities
Construction engineering										Emergency support facilities
Transportation facilities										Lifeline engineering facilities
Transportation facilities										Emergency support facilities
Lifeline engineering facilities										Emergency support facilities

What is the relative importance of each of the following sets of pairwise comparisons of elements to "urban management safety resilience?"

X	Importance comparison									Y
	1/5	1/4	1/3	1/2	1	2	3	4	5	
Risk control level										Support and guarantee input level
Risk control level										Material supply level
Support and guarantee input level										Material supply level

What is the relative importance of each of the following sets of pairwise comparisons of elements to "basic population attributes?"

X	Importance comparison									Y
	1/5	1/4	1/3	1/2	1	2	3	4	5	
The density of permanent residents in the built-up area										The level of basic medical insurance for urban workers
The density of permanent residents in the built-up area										The proportion of temporary residents
The density of permanent residents in the built-up area										The number of newly added higher education talents
The level of basic medical insurance for urban workers										The proportion of temporary residents
The level of basic medical insurance for urban workers										The number of newly added higher education talents
The proportion of temporary residents										The number of newly added higher education talents

What is the relative importance of each of the following sets of pairwise comparisons of elements to "social participation readiness?"

X	Importance comparison									Y
	1/5	1/4	1/3	1/2	1	2	3	4	5	
The level of reserve health technical personnel in the city										The reserve level of urban health institutions
The level of reserve health technical personnel in the city										The level of social organization units
The reserve level of urban health institutions										The level of social organization units

What is the relative importance of each of the following sets of pairwise comparisons of elements to "safety perception and safety culture?"

X	Importance comparison									Y
	1/5	1/4	1/3	1/2	1	2	3	4	5	
The number of people covered by unemployment insurance										The number of people covered by urban old-age insurance
The number of people covered by unemployment insurance										The number of people covered by work-related injury insurance
The number of people covered by unemployment insurance										The per capita emergency deposit level
The number of people covered by urban old-age insurance										The number of people covered by work-related injury insurance
The number of people covered by urban old-age insurance										The per capita emergency deposit level
The number of people covered by work-related injury insurance										The per capita emergency deposit level

What is the relative importance of each of the following sets of pairwise comparisons of elements to "construction engineering?"

X	Importance comparison									Y
	1/5	1/4	1/3	1/2	1	2	3	4	5	
Urbanization rate										Proportion of land area in safety-vulnerable areas
Urbanization rate										The level of personnel reserves in the construction industry
Proportion of land area in safety-vulnerable areas										The level of personnel reserves in the construction industry

What is the relative importance of each of the following sets of pairwise comparisons of elements to "transportation facilities?"

X	Importance comparison									Y
	1/5	1/4	1/3	1/2	1	2	3	4	5	
Per capita urban road area										The level of urban traffic lighting facilities

What is the relative importance of each of the following sets of pairwise comparisons of elements to "lifeline engineering facilities?"

X	Importance comparison									Y
	1/5	1/4	1/3	1/2	1	2	3	4	5	
The penetration rate of wired broadcasting and television										The level of urban drainage facilities

What is the relative importance of each of the following sets of pairwise comparisons of elements to "emergency support facilities?"

X	Importance comparison									Y
	1/5	1/4	1/3	1/2	1	2	3	4	5	
Per capita refuge area										Building area of disaster relief reserve institution warehouses per ten thousand people
Per capita refuge area										The number of beds in medical and health institutions per ten thousand people
Per capita refuge area										The green coverage rate
Building area of disaster relief reserve institution warehouses per ten thousand people										The number of beds in medical and health institutions per ten thousand people
Building area of disaster relief reserve institution warehouses per ten thousand people										The green coverage rate
The number of beds in medical and health institutions per ten thousand people										The green coverage rate

What is the relative importance of each of the following sets of pairwise comparisons of elements to "risk control level?"

X	Importance comparison									Y
	1/5	1/4	1/3	1/2	1	2	3	4	5	
The death rate per million people due to disasters										The proportion of annual direct economic losses due to disasters to the region's gross domestic product

What is the relative importance of each of the following sets of pairwise comparisons of elements to "support and guarantee input level?"

X	Importance comparison									Y
	1/5	1/4	1/3	1/2	1	2	3	4	5	
The level of financial support for public safety										The level of financial support for medical and health care
The level of financial support for public safety										The level of financial support for medical and health care
The level of financial support for medical and health care										The level of financial support for transportation

What is the relative importance of each of the following sets of pairwise comparisons of elements to "material supply level?"

X	Importance comparison									Y
	1/5	1/4	1/3	1/2	1	2	3	4	5	
The comprehensive production capacity of water supply										The level of grain supply
The comprehensive production capacity of water supply										The level of emergency material supply
The level of grain supply										The level of emergency material supply